METAL CUTTING

PREFACE

I have been fortunate in having many able and congenial colleagues during the twenty-five years in which I have been actively interested in the subject of metal cutting. Without their collaboration and the contributions which they have made in skill and ideas this book would not have been written. I would like here to acknowledge the part which they have played in developing understanding of the metal cutting process.

My ideas on the subject were formulated during the years when I was employed by Hard Metal Tools Ltd. of Coventry (now Wickman–Wimet Ltd.) in research and development on cemented carbide tool materials. In the last eight years in the Industrial Metallurgy Department at the University of Birmingham, work with a group of technicians, staff and students has helped to develop these ideas further and to broaden the subject matter. If for no other reason than the length of Chapter six, on the subject of tool materials, the reader will hardly fail to notice the central position in this picture of metal cutting which is occupied by the tool. This emphasis must be attributed very largely to my personal involvement in metal cutting as a metallurgist whose major interest has been in tool materials. However, it is not inappropriate to look at metal cutting from this point of view.

In his book *Man the Tool Maker*, published by the British Museum, Kenneth P. Oakley states the proposition 'Human progress has gone step by step with the discovery of better materials of which to make *cutting* tools, and the history of man is therefore broadly divisible into the Stone Age, the Bronze Age, the Iron Age and the Steel Age'. Certainly, in the last 100 years, since the alloy tool steels of Robert Mushet, the accelerated development of new materials for cutting metals has made the most important contribution to the vast increase in machining efficiency which has characterised engineering industry in this period. This is because the ability of the tool to withstand the stresses and temperatures involved has been the factor limiting the speed of cutting and the forces and power which could be employed. Unfortunately the key role played by tool materials has not been appreciated by industrial and political leaders in the United Kingdom. Very important sectors of tool material production have passed into the hands of large corporations controlled from outside the country. This is in spite of the fact that in this relatively small, vital sector of industry we have at present, as in the past, the technical and scientific expertise to maintain a position of excellence at least equal to any in the world.

The metallurgical emphasis of this book is largely due to my time as a student at Sheffield University in the Metallurgy Department where two people in particular, Dr Edwin Gregory and Mr G.A. de Belin were responsible for a very high level of teaching in the techniques and interpretations of metallography. The most important element in the evidence presented here comes from optical metallography. The field of useful observation is being greatly extended by electron microscopy and by instruments such as the microprobe analyser, but a high level of optical metallography remains at the centre of such metallurgical investigations. Studies of the significant regions in metal cutting present many technical difficulties for the metallographer and I would like to pay tribute to the skill and painstaking work of many of my colleagues, and in particular to Mr E.F. Smart, who has worked continuously in this region for many years and has taken many of the photomicrographs presented here. Dr P.K. Wright, Dr B. Dines and Dr R. Freeman at the University of Birmingham, and many of the staff at Wickman–Wimet Laboratories, have contributed to the metallurgical observations.

Ideas in a complex area such as metal cutting are shaped by innumerable contributions varying from major research projects to informal discussions. It is impossible to make adequate reference to all the participants, but apart from colleagues already mentioned, I would like to acknowledge in particular the contributions of Mr E. Lardner, Dr D.R. Milner, and Professor G.W. Rowe. In this work I have had available the resources of Wickman–Wimet Ltd. under the former Research Director, Mr A.E. Oliver, and of the Industrial Metallurgy Department at the University of Birmingham under the late Professor E.C. Rollason and Professor D.V. Wilson.

Not all of my colleagues would agree with all the notions incorporated in this picture of metal cutting. I accept responsibility for the general philosophy of the book and for the detailed form in which it is presented. If the result is to help some of those engaged in practice in machining to take more rational decisions because of better understanding of what happens during metal cutting, or if further research in this area is stimulated then I will consider that the effort of putting the book together has been worthwhile.

My wife has contributed in innumerable ways to the writing of the book, but especially by spending many hours carefully reading the typescript, correcting and improving the presentation. I am very grateful to her for this assistance and for her patience during the months that I have been preoccupied with this work.

E.M. Trent

THE BUTTERWORTH GROUP

United Kingdom	**Butterworth & Co (Publishers) Ltd** London: 88 Kingsway, WC2B 6AB
Australia	**Butterworths Pty Ltd** Sydney: 586 Pacific Highway, Chatswood, NSW 2067 Also at Melbourne, Brisbane, Adelaide and Perth
Canada	**Butterworth & Co (Canada) Ltd** Toronto: 2265 Midland Avenue, Scarborough, Ontario, M1P 4S1
New Zealand	**Butterworths of New Zealand Ltd** Wellington: T & W Young Building, 77–85 Customhouse Quay, 1, CPO Box 472
South Africa	**Butterworth & Co (South Africa) (Pty) Ltd** Durban: 152–154 Gale Street
USA	**Butterworths (Publishers) Inc** Boston: 19 Cummings Park, Woburn, Mass 01801

First published 1977
Reprinted 1978

ISBN 0 408 10603 4

© Butterworth & Co (Publishers) Ltd, 1977

Library of Congress Cataloging in Publication Data

Trent, Edward Moor.
  Metal cutting.

  Bibliography: p.
  Includes index.
  1. Metal-cutting. 2. Metal-cutting tools.
3. Tool-steel. I. Title.
TJ1185.T73   671.5'3   76–16542
ISBN 0-408-10603-4

Printed in Great Britain by Butler & Tanner Ltd., Frome and London

Metal Cutting

E.M. TRENT, Ph.D., D. Met., F.I.M.
Department of Industrial Metallurgy, University of Birmingham

BUTTERWORTHS
LONDON – BOSTON
Sydney – **Wellington** – Durban – Toronto

ACKNOWLEDGEMENT

Many of the illustrations have been published previously. The Author wishes to thank the following for permission to reproduce the illustrations listed.

The Metals Society – Figures 3.5, 3.7, 3.9, 3.15, 3.16, 5.3, 5.4, 5.8a, 6.3, 6.6, 6.8, 6.9, 6.10, 6.11, 6.12, 6.14, 6.16, 6.17, 6.20, 6.30, 6.35, 6.37, 6.40, 6.44, 6.51, 7.2, 7.11, 7.19, 7.23, 7.27, 7.32, 8.8, 8.9, 8.10, 8.13, 8.14.

International Journal for Production Research – Figures 5.5, 5.6, 5.7, 5.10, 7.28, 7.29, 7.30.

International Journal for Machine Tool Design and Research – Figures 7.6, 8.1, 8.2, 8.3, 8.4, 8.5.

American Society for Metals – Figures 6.21, 7.17.

The Author also wishes to thank Wickman–Wimet Ltd for permission to reproduce the following photomicrographs, taken in their laboratories, some of which have not been previously published: Figures 3.5, 3.7, 3.8, 3.9, 3.15, 6.23, 6.27, 6.29, 6.30, 6.31, 6.35, 6.36, 6.37, 6.38, 6.39, 6.40, 6.41, 6.42, 6.43, 6.44, 6.45, 6.46, 6.47, 6.48, 6.49, 6.51, 6.52, 6.53, 6.54, 7.2, 7.6, 7.10, 7.18, 7.19, 7.22, 7.23, 7.26, 7.27, 7.32, 8.14.

The Author is grateful to his colleagues Dr E. Amini, Dr P.K. Wright, Dr R.M. Freeman and Dr B.W. Dines for photomicrographs and graphs acknowledged in the captions to the illustrations which they have contributed.

CONTENTS

ANSI

1

INTRODUCTION

Subject matter

It would serve no useful purpose to attempt a precise definition of metal cutting. In this book the term is intended to include operations in which a thin layer of metal, the *chip* or *swarf*, is removed by a wedge-shaped tool from a larger body. There is no hard and fast line separating *chip-forming* operations from others such as the shearing of sheet metal, the punching of holes or the cropping of lengths from a bar. These also can be considered as metal cutting, but the action of the tools and the process of separation into two parts are so different from those encountered in chip-forming operations, that the subject requires a different treatment and these operations are not considered here.

There is a great similarity between the operations of cutting and grinding. Our ancestors ground stone tools before metals were discovered and later used the same process for sharpening metal tools and weapons. The grinding wheel does much the same job as the file, which can be classified as a cutting tool, but has a much larger number of cutting edges, randomly shaped and oriented. Each edge removes a much smaller fragment of metal than is normal in cutting, and it is largely because of this difference in size that conclusions drawn from investigations into metal cutting must be applied with reservations to the operation of grinding.

In the engineering industry, the term *machining* is used to cover chip-forming operations, and a definition with this meaning appears in many dictionaries. Most machining is carried out to shape metals and alloys, but the lathe was first used to turn wood and bone, and today many plastic products also are machined. The term metal cutting is used here because research has shown certain characteristic features of the behaviour of metals during cutting which dominate the process, and without further work, it is not possible to extend the principles described here to embrace the cutting of other materials.

Historical

Before the middle of the 18th century the main material used in engineering structures was wood. The lathe and a few other machine tools

1

existed, mostly constructed in wood and most commonly used for shaping wooden parts. The boring of cannon and the production of metal screws and small instrument parts were the exceptions. It was the steam engine, with its requirement of large metal cylinders and other parts of unprecedented dimensional accuracy, which led to the first major developments in metal cutting.

The materials of which the first steam engines were constructed were not very difficult to machine. Grey cast iron, wrought iron, brass and bronze were readily cut using hardened carbon steel tools. The methods of heat treatment of tool steel had been evolved by centuries of craftsmen, and reasonably reliable tools were available, although rapid failure of the tools could be avoided only by cutting very slowly. It required 27½ working days to bore and face one of Watt's large cylinders.[1]

At the inception of the steam engine, no machine tool industry existed – the whole of this industry is the product of the last two hundred years. The century from 1760 to 1860 saw the establishment of enterprises devoted to the production of machine tools. Maudslay, Whitworth, and Eli Whitney, among many other great engineers, devoted their lives to perfecting the basic types of machine tools required for generating, in metallic components, the cylindrical and flat surfaces, threads, grooves, slots and holes of many shapes required by the developing industries.[1] The lathe, the planer, the shaper, the milling machine, drilling machines and power saws were all developed into rigid machines capable, in the hands of good craftsmen, of turning out large numbers of very accurate parts and structures of sizes and shapes that had never before been contemplated. By 1860 the basic problems of how to produce the necessary shapes in the existing materials had largely been solved. There had been little change in the materials which had to be machined – cast iron, wrought iron and a few copper based alloys. High carbon tool steel, hardened and tempered by the blacksmith, still had to answer all the tooling requirements. The quality and consistency of tool steels had been greatly improved by a century of experience with the crucible steel process, but the limitations of carbon steel tools at their best were becoming an obvious constraint on speeds of production.

From 1860 to the present day, the emphasis has shifted from development of the basic machine tools and the know-how of production of the required shapes and accuracy, to the problems of machining new metals and alloys and to the reduction of machining costs. With the Bessemer and Open Hearth steel making processes, steel rapidly replaced wrought iron as a major construction material. The tonnage of steel soon vastly exceeded the earlier output of wrought iron and much of this had to be machined. Steel proved more difficult to machine than wrought iron, and cutting speeds had to be lowered to achieve a reasonable tool life. When cutting alloy steels speeds were even lower, and towards the end of the 19th century the costs of machining were becoming very great in terms of manpower and capital investment. The incentive to reduce costs by cutting faster and automating the cutting process became more intense, and, up to the present time, continues to be the mainspring of the major developments in the metal cutting field.

The technology of metal cutting has been improved by contributions from all the branches of industry with an interest in machining. The replacement of carbon tool steel by high speed steels and cemented carbides has allowed cutting speeds to be increased by many times. Machine tool manufacturers have developed machines capable of making full use of the new tool materials, while automatic machines, numerically controlled machines and transfer machines greatly increase the output per worker employed. Tool designers and machinists have optimised the shapes of tools to give long tool life at high cutting speed. Lubricant manufacturers have developed many new coolants and lubricants to improve surface finish and permit increased rates of metal removal.

The producers of those metallic materials which have to be machined played a double role. Many new alloys were developed to meet the increasingly severe conditions of stress, temperature and corrosive atmosphere imposed by the requirements of our industrial civilisation. Some of these, like aluminium and magnesium, are easy to machine, but others, like high-alloy steels and nickel-based alloys, become more difficult to cut as their useful properties improve. On the other hand, metal producers have responded to the demands of production engineers for metals which can be cut faster. New heat treatments have been devised, and the introduction of alloys like the free-machining steels and leaded brass has made great savings in production costs.

Today metal-cutting is a very large segment indeed of our industry. The motor car industry, electrical engineering, railways, shipbuilding, aircraft manufacture, production of domestic equipment and the machine tool industry itself — all these have large machine shops with many thousands of employees engaged in machining. In 1971 in the UK there were over 1 million machine tools, 85% of which were metal cutting machines.[2] There are thus more than a million people directly engaged in machining and others in subsidiary activities. The cost of machining in the USA in 1971 was calculated to be approximately $40 000 m.[3] In 1957 an estimate showed that more than 15 million tons of metal were turned into swarf in the USA,[4] and in the same year 100 million tons of steel were produced. This suggests that something like 10% of all the metal produced is turned into chips.

Thus metal cutting is a very major industrial activity, employing tens of millions of people throughout the world. The wastefulness of turning so much metal into low grade scrap has directed attention to methods of reducing this loss. Much effort has been devoted to the development of ways of shaping components in which metal losses are reduced to a minimum — employing processes such as cold-forging, precision casting and powder metallurgy — and some success has been achieved. However, the weight of components cold forged from steel in 1970 in the UK was less than 50 thousand tons (excluding cold-headed bolts), and the annual production of powder metallurgy products is of the order of 15 thousand tons. So far there are few signs that the numbers of people employed or the money expended on machining are being significantly reduced. In spite of its evident wastefulness, it is still the cheapest way to make very many shapes and is likely to continue to be so for many years. The further evolution of the technology of machining to higher standards of efficiency and accuracy, and with less intolerable working conditions, is of great importance to industry generally.

Progress in the technology of machining is achieved by the ingenuity and experiment, the intuition, logical thought and dogged worrying of many thousands of practitioners engaged in the many-sided arts of metal cutting. The worker operating the machine, the tool designer, the lubrication engineer, the metallurgist, are all constantly probing to find answers to new problems in the form of novel materials, limitations on the speed of production or the precision achievable and many others.   However competent they may be, there can be few craftsmen, technologists or scientists engaged in this field who do not feel that they would be better able to solve their problems if they had a deeper knowledge of what was happening at the cutting edge of the tool.

It is what happens in a very small volume of metal around the cutting edge that determines the performance of tools, the machinability of metals and alloys and the qualities of the machined surface.   During cutting, the interface between tool and work material is largely inaccessible to observation, but indirect evidence concerning stresses, temperatures, metal flow and interactions between tool and work material has been contributed by many researches.   This book will endeavour to summarise the available knowledge concerning this region derived from published work, the Author's own research and that of many of his colleagues.

References

1.   ROLT, L.T.C.   *Tools for the Job,* Batsford (1965)
2.   COOKSON, J.O. and SWEENEY, G.   I.S.I. Publication No.138, 83 (1971)
3.   ZLATIN, N. *1st. Int. Cemented Carbide Conf., Chicago,* paper No.IQ7l-918
4.   MERCHANT, M.E. *Trans. Amer. Soc. Mech. Eng.,* **79,** 1137 (1957)

2

METAL CUTTING OPERATIONS AND TERMINOLOGY

Of all the processes used to shape metals, it is in machining that the conditions of operation are most varied. Almost all metals and alloys are machined — hard or soft, cast or wrought, ductile or brittle, with high or low melting point. Most shapes used in the engineering world are sometimes produced by machining and, as regards size, components from watch parts to pressure vessels more than 3 m (10 ft) in diameter are shaped by cutting tools. Many different machining operations are used, involving cutting speeds as high as 600 m min^{-1} (2 000 ft/min) or as low as a few centimetres (inches) per minute, while cutting may be continuous for several hours or interrupted in fractions of a second.

It is important that this great variability be appreciated. Some of the major variables are, therefore, discussed and the more significant terms defined by describing briefly some of the more important metal cutting operations with their distinctive features.

Turning

This basic operation is also the one most commonly employed in experimental work on metal cutting. The work material is held in the chuck of a lathe and rotated. The tool is held rigidly in a tool post and moved at a constant rate along the axis of the bar, cutting away a layer of metal to form a cylinder or a surface of more complex profile. This is shown diagrammatically in *Figure 2.1*.

The *cutting speed* (*V*) is the rate at which the uncut surface of the work passes the cutting edge of the tool — usually expressed in units of ft/min or m min^{-1}. The *feed rate* (*f*) is the distance moved by the tool in an axial direction at each revolution of the work, *Figure 2.1c*. The *depth of cut* (*w*) is the thickness of metal removed from the bar, measured in a radial direction, *Figure 2.1b & c*. The product of these three gives the rate of metal removal, a parameter often used in measuring the efficiency of a cutting operation.

$$V.f.w. \quad = \quad \text{rate of metal removal}$$

The cutting speed and the feed rate are the two most important parameters which can be adjusted by the operator to achieve optimum cutting conditions. The depth of cut is often fixed by the initial size of the bar and the required size of the product. Cutting speed is usually between 3

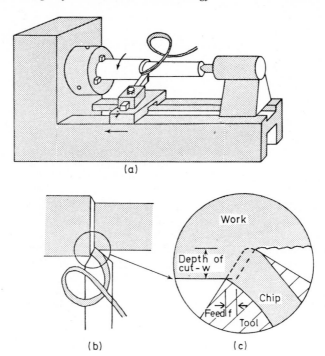

(a)

(b)   (c)

Figure 2.1   Lathe turning

and 200 m min⁻¹ (10 and 600 ft/min) but in exceptional cases, may be
as high as 3 000 m min⁻¹ (10 000 ft/min).   Feed rates may be as low as
0.0125 mm (0.0005 in) per rev. and with very heavy cutting up to
2.5 mm (0.1 in) per rev. Depth of cut may vary from nil over part of
the cycle to over 25 mm (1 in). It is possible to remove metal at a rate
of more than 1 600 cm³ (100 in³) per minute, but such a rate would be
very uncommon and 80-160 cm³ (5-10 in³) per minute would normally
be considered as rapid.

   Figure 2.2 shows some of the main features of a turning tool.   The
surface of the tool over which the chip flows is known as the *rake face.*
The *cutting edge* is formed by the intersection of the rake face with the
clearance face or *flank* of the tool.   The tool is so designed and held in
such a position that the clearance face does not rub against the freshly
cut metal surface.   The *clearance angle* is variable but is often of the
order of 6-10°.   The rake face is inclined at an angle to the axis of the
bar of work material and this angle can be adjusted to achieve optimum
cutting performance for particular tool materials, work materials and
cutting conditions.   The *rake angle* is measured from a horizontal line.
A *positive rake* angle is one where the rake face dips below the line,
Figure 2.2c, and early metal cutting tools had large positive rake angles
to give a cutting edge which was keen, but easily damaged.   Positive rake

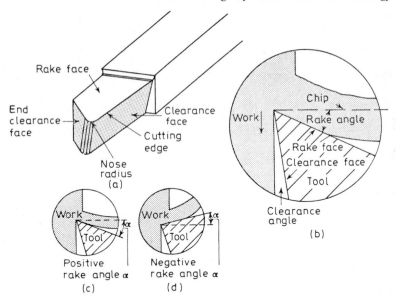

Figure 2.2 Cutting tool terminology

angles may be up to 30° but the greater robustness of tools with smaller
rake angle leads in many cases to the use of zero or *negative rake* angle,
Figure 2.2d. With a negative rake angle of 5° or 6°, the included angle
between the rake and clearance faces may be 90°, and this has advantages.

The tool terminates in an *end clearance face, Figure 2.2a,* which also
is inclined at such an angle as to avoid rubbing against the freshly cut
surface. The *nose* of the tool is at the intersection of all three faces
and may be sharp, but more frequently there is a *nose radius* between
the two clearance faces.

This very simplified description of the geometry of one form of turning tool is intended to help the reader without practical experience of
cutting to follow the terms used later in the book. The design of tools
involves an immense variety of shapes and the full nomenclature and
specifications are very complex.[1,2] It is difficult to appreciate the action
of many types of tool without actually observing or, preferably, using
them. The performance of cutting tools is very dependent on their precise shape. In most cases there are critical features or dimensions which
must be accurately formed for efficient cutting. These may be, for example, the clearance angles, the nose radius and its blending into the faces,
or the sharpness of the cutting edge. The importance of precision in
tool making, whether in the tool room of the user, or in the factory of
the tool maker, cannot be over estimated. This is an area where excellence in craftsmanship is still of great value.

Having considered turning, a number of other machining operations are
discussed briefly, with the object of demonstrating some characteristic
differences and similarities in the cutting parameters.

Boring

Essentially the conditions of boring of internal surfaces differ little from those of turning, but this operation illustrates the importance of rigidity in machining. Particularly when a long cylinder is bored, the bar holding the tool must be long and slender and cannot be as rigid as the thick, stocky tools and tool post used for most turning. The tool tends to be deflected to a greater extent by the cutting forces and to vibrate. Vibration may affect not only the dimensions of the machined surface, but its roughness and the life of the cutting tools.

Drilling

In drilling, carried out on a lathe or a drilling machine, the tool most commonly used is the familiar twist-drill. The 'business end' of a twist drill has two cutting edges. The rake faces of the drill are formed by part of each of the *flutes, Figure 2.3,* the rake angle being the *helix angle* of the drill. The chips slide up the flutes, while the end faces must be ground at the correct angle to form the clearance face.

An essential feature of drilling is the variation in cutting speed along the cutting edge. The speed is a maximum at the periphery, which generates the cylindrical surface, and approaches zero near the centre-line of the drill, the *web, Figure 2.3,* where the cutting edge is blended to a chisel shape, and the action is no longer that of a cutting tool in the same sense. This variation in speed along the edge is responsible for many aspects of drilling which are peculiar to this operation.

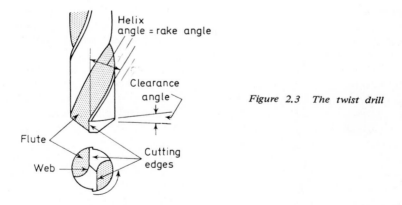

Figure 2.3 The twist drill

Facing

Turning, boring and drilling generate cylindrical or more complex surfaces of rotation. Facing, also carried out on a lathe generates a flat surface, normal to the axis of rotation, by feeding the tool from the surface towards the centre or outward from the centre. In facing, the depth of cut is measured in a direction parallel to the axis and the feed in a radial direction. A characteristic of this operation is that the cutting speed varies continuously, approaching zero towards the centre of the bar.

Forming and parting off

Surfaces of rotation of complex form may be generated by turning, but some shapes may be formed more efficiently by use of a tool with a cutting edge of the required profile, which is fed into the peripheral surface of the bar in a radial direction. Such tools often have a long cutting edge. The part which touches the tool first cuts for the longest time and makes the deepest part of the form, while another part of the tool is cutting for only a very short part of the cycle. Cutting forces, with such a long edge, may be high and feed rates are usually low.

Many short components, such as spark plug bodies, are made from a length of bar, the final operation being parting off, a thin tool being fed into the bar from the periphery to the centre or to a central hole. The tool must be thin to avoid waste of material, but it may have to penetrate to a depth of several centimetres. These slender tools must not be further weakened by large clearance angles down the sides. Both parting and forming tools have special difficulties associated with small clearance angles, which cause rapid wear at localised positions.

Milling

Both grooves and flat surfaces – for example the faces of a car cylinder block – are generated by milling. In this operation the cutting action is achieved by rotating the tool, while the work is held on a table and the feed action is obtained by moving it under the cutter, *Figure 2.4*. There

Figure 2.4 Milling cutters

is a very large number of different shapes of milling cutter for different applications. Single toothed cutters are possible but typical milling cutters have a number of teeth (cutting edges) which may vary from three to over one hundred. The new surface is generated as each tooth cuts away an arc-shaped segment, the thickness of which is the *feed* or *tooth load*. Feeds are usually light, not often greater than 0.25 mm (0.01 in) per tooth, and frequently as low as 0.025 mm (0.001 in) per tooth. However, because of the large number of teeth, the rate of metal removal is often high.

An important feature of all milling operations is that the action of each cutting edge is intermittent. Each tooth is cutting during less than half of a revolution of the cutter, and sometimes for only a very small part of the cycle. Each edge is subjected to periodic impacts as it makes contact with the work, is stressed and heated during the cutting part of the cycle, followed by a period when it is unstressed and allowed to cool. Frequently cutting times are a small fraction of a second and are repeated several times a second, involving both thermal and mechanical fatigue of the tool. The design of milling cutters is greatly influenced by the problem of getting rid of the chips so that they do not interfere with the cutting action.

Milling is used also for the production of curved shapes, while end mills, which are larger and more robust versions of the dentist's drill, are employed in the production of hollow shapes such as die cavities.

Shaping and planing

These are two other methods for generating flat surfaces and can also be used for producing grooves and slots. In shaping, the tool has a reciprocating movement, the cutting taking place on the forward stroke along the whole length of the surface being generated, while the reverse stroke is made with the tool lifted clear, to avoid damage to the tool or the work. The next stroke is made when the work has been moved by the feed distance, which may be either horizontal or vertical. Planing is similar, but the tool is stationary, and the cutting action is achieved by moving the work. The reciprocating movement, involving periodic reversal of large weights, means that cutting speeds are relatively slow, but fairly high rates of metal removal are achieved by using high feed rates. The intermittent cutting involves severe impact loading of the cutting edge at every stroke. The cutting times between interruptions are longer than in milling but shorter than in most turning operations.

The operations described are some of the more important ones employed in shaping engineering components, but there are many others. Some of these, like sawing, filing and tapping of threads, are familiar to all of us, but others such as broaching, skiving, reaming or the hobbing of gears are the province of the specialist. Each of these operations has its special characteristics and problems, as well as the features which it has in common with the others.

There is one further variable which should be mentioned at this stage and this is the environment of the tool edge. Cutting is often carried out in air, but, in many operations, the use of a fluid to cool the tool or the workpiece, and/or act as a lubricant is essential to efficiency. The fluid is usually a liquid, based on water or a mineral oil, but may be a gas such as CO_2. The action of coolants/lubricants will be considered in the last chapter, but the influence of the cutting lubricant, if any, should never be neglected.

One object of this discussion of cutting operations is to make a major point which the reader should bear in mind in the subsequent chapters. In spite of the complexity and diversity of machining in the hard world of the engineering industry, an analysis is made of some of the more important features which metal cutting operations have in common. The work on which this analysis is based has, of necessity, been derived from a study of a limited number of cutting operations — often the simplest — such as uninterrupted turning, and from investigations in machining a limited range of work materials. The results of this analysis should not be used uncritically, because, in the complex conditions of machine shops, there can be relatively few statements which apply universally. In considering any problem of metal cutting, the first questions to be asked should be — what is the machining operation, and what are the specific features which are critical for this operation?[3]

References

1. British Standard 1886:
2. STABLER, G.V. *J.I. Prod. E.,* **34,** 264 (1955)
3. *A.S.M. Metals Handbook,* 8th ed., Vol.3. 'Machining' (1967)

3

THE ESSENTIAL FEATURES OF METAL CUTTING

The processes of cutting most familiar to us in everyday life are those in which very soft bodies are severed by a tool such as a knife, in which the cutting edge is formed by two faces meeting at a very small included angle. The wedge-shaped tool is forced symmetrically into the body being cut, and often, at the same time, is moved parallel with the edge, as when slicing bread. If the tool is sharp, the body may be cut cleanly, with very little force, into two pieces which are gently forced apart by the faces of the tool. A microtome will cut very thin layers from biological specimens with no observable damage to the layer or to the newly-formed surface.

Metal cutting is not like this. Metals and alloys are too hard, so that no known tool materials are strong enough to withstand the stresses which they impose on very 'fine' cutting edges. (Very low melting point metals, such as lead and tin, may be cut in this way, but this is exceptional). If both faces forming the tool edge act to force apart the two newly-formed surfaces, very high stresses are imposed, much heat is generated, and both the tool and the work surfaces are damaged. These considerations make it necessary for a metal-cutting tool to **take the form of a large-angled wedge, which is driven asymmetrically into the work material, to remove a thin layer from a thicker body.** (*Figure 3.1*). The layer must be thin to enable the tool and work to withstand the imposed stress, and a clearance angle must be formed on the tool to ensure that the clearance face does not make contact with the newly-formed work surface. In spite of the diversity of machining operations, emphasised in Chapter 2, these are features of all metal cutting operations and provide common ground from which to commence an analysis of machining.

In practical machining, the included angle of the tool edge varies between 55° and 90°, so that the removed layer, the chip, is diverted through an angle of at least 60° as it moves away from the work, across the rake face of the tool. In this process, **the whole volume of metal removed is plastically deformed**, and thus a large amount of energy is required to form the chip and to move it across the tool face. In the process, two new surfaces are formed, the new surface of the work-piece (*OA* in *Figure 3.1*) and the under surface of the chip (*BC*). The formation of new surfaces requires energy, but in metal cutting, the theoretical minimum energy required to form the new surfaces is an insignificant proportion of that required to deform plastically the whole of the metal removed.

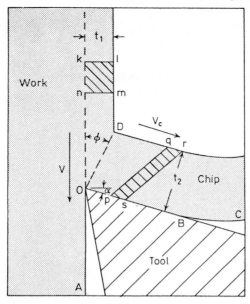

Figure 3.1 Metal cutting diagram

The main objective of machining is the shaping of the new work surface. It may seem, therefore, that too much attention is paid in this
book to the formation of the chip, which is a waste product. But the
consumption of energy occurs mainly in the formation and movement of
the chip, and for this reason, the main economic and practical problems
concerned with rate of metal removal and tool performance, can be understood only by studying the behaviour of the work material as it is formed
into the chip and moves over the tool. Even the condition of the
machined surface itself can be understood only with knowledge of the process of chip formation.

The chip

The chip is enormously variable in shape and size in industrial machining
operations; *Figure 3.2* shows some of the forms. The formation of all
types of chip involves a shearing of the work material in the region of a
plane extending from the tool edge to the position where the upper surface of the chip leaves the work surface (*OD* in *Figure 3.1*). A very
large amount of strain takes place in this region in a very short interval
of time, and not all metals and alloys can withstand this strain without
fracture. Grey cast iron chips, for example, are always fragmented, and
the chips of more ductile materials may be produced as segments under
unfavourable conditions of cutting. This *discontinuous chip* is one of the
principal classes of chip form, and has the practical advantage that it is

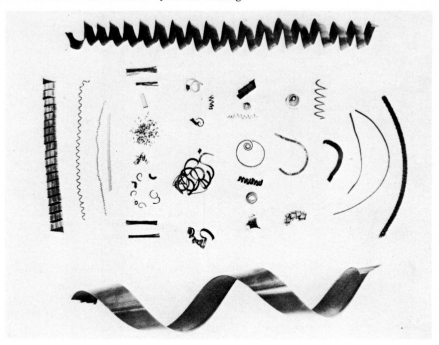

Figure 3.2   Chip shapes

easily cleared from the cutting area.   Under a majority of cutting conditions, however, ductile metals and alloys do not fracture on the shear plane and a *continuous chip* is produced, *Figure 3.3*.   Continuous chips may adopt many shapes — straight, tangled or with different types of helix.   Often they have considerable strength, and control of chip shape is one of the problems confronting machinists and tool designers.   Continuous and discontinuous chips are not two sharply defined categories, and every shade of gradation between the two types can be observed.

The longitudinal shape of continuous chips can be modified by mechanical means, for example by grooves in the tool rake face, which curl the chip into a helix.   The cross section of the chips and their thickness are of great importance in the analysis of metal cutting, and are considered in some detail.   For the purpose of studying chip formation in relation to the basic principles of metal cutting, it is useful to start with the simplest possible cutting conditions, consistent with maintaining the essential features common to these operations.

TECHNIQUES FOR STUDY OF CHIP FORMATION

Before discussing chip shape, the experimental methods used for gathering the information are described.   The simplified conditions used in the first stages of laboratory investigations are known as *orthogonal* cutting.   In orthogonal cutting the tool edge is straight, it is normal to the direction of cutting, and normal also to the feed direction. On a lathe, these conditions are secured by using a tool with the cutting edge horizontal, on the centre line, and at right angles to the axis of rotation of the workpiece. If the workpiece is in the form of a tube, whose wall thickness is the depth of cut, only the straight edge of the tool is used. In this method the cutting speed is not quite constant along the cutting edge, being highest at the outside of the tube, but if the tube diameter is reasonably large this is of minor importance. In many cases the work material is not available in tube form, and what is sometimes called *semi-orthogonal* cutting conditions are used, in which the tool cuts a solid bar with a constant depth of cut. In this case, conditions at the nose of the tool are different from those at the outer surface of the bar. If a sharp-nosed tool is used this may result in premature failure, so that it is more usual to have a small nose radius. To avoid too great a departure from orthogonal conditions, the major part of the edge engaged in cutting should be straight.

Strictly orthogonal cutting can be carried out on a planing or shaping machine, in which the work material is in the form of a plate, the edge of which is machined. The cutting action on a shaper is, however, intermittent, the time of continuous machining is very short, and speeds are limited. For most test purposes the lathe method is more convenient.

The study of the formation of chips is difficult, because of the high speed at which it takes place under industrial machining conditions, and the small scale of the phenomena which are to be observed. High speed cine-photography at low magnification has been used to reveal the changing external shape. This can give an impression of cutting action not possible by other methods, but live cine-projection is the only satisfactory means of presenting the information gained. Other limitations are that the depth of focus and the definition are inadequate for observation of fine details, and that only action at the external surface can be observed. As shown later, this may be misleading if interpreted as demonstrating the cutting action at the centre of the chip.

No useful information about chip formation can be gained by studying the end of the cutting path after cutting has been stopped in the normal way by disengaging the feed and the drive to the work. By stopping the cutting action suddenly, however, it is possible to retain many of the important details — to 'freeze' the action of cutting. Several 'quick-stop' mechanisms have been devised for this purpose. One of the most successful involves the use of a humane killer gun to propel a lathe tool away from the cutting position at very high speed in the direction in which the work material is moving.[1] Sometimes the chip adheres to the tool and separates from the bar, but more usually the tool comes away more or less cleanly, leaving the chip attached to the bar. A segment of the bar, with chip attached, can be cut out and examined in detail at any required magnification. For external examination and photography, the scanning electron microscope (SEM) is particularly valuable because of its great depth of focus. *Figure 3.3* shows an example of chip formation recorded in this way.

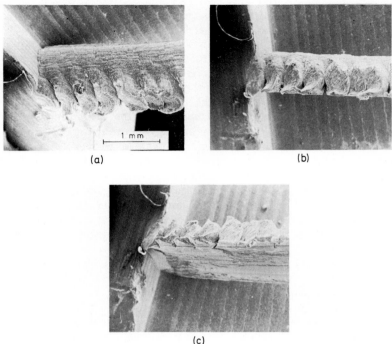

(a)   (b)

(c)

Figure 3.3   SEM photographs from three directions of a forming chip of steel cut at 48 m min^{-1} (150 ft/min) and 0.25 mm (0.01 in) per rev feed. (B. W. Dines)

Much of the information in this book has been obtained by preparing metallographic sections through 'quick-stop' sections to reveal the internal action of cutting.   Because any one specimen illustrates the cutting action at one instant of time, several specimens must be prepared to distinguish those features which have general significance from others which are peculiar to the instant at which cutting stopped.

CHIP SHAPE

Even with orthogonal cutting, the cross section of the chip is not strictly rectangular.   Since it is constrained only by the rake face of the tool, the metal is free to move in all other directions as it is formed into the chip. The chip tends to spread sideways, so that the maximum width is somewhat greater than the original depth of cut.   In cutting a tube it can spread in both directions, but in turning a bar it can spread only outward.   The chip spread is small with harder alloys, but when cutting soft metals with a small rake angle tool, a chip width more than one and one half times the depth of cut has been observed.   Usually the chip *thickness* is greatest near the middle, tapering off somewhat towards the sides.

The upper surface of the chip is always rough, usually with minute corrugations or steps, *Figure 3.3*.   Even with a strong, continuous chip, periodic cracks are often observed, breaking up the outer edge into a series of segments.   A complete description of chip form would be very complex, but, for the purposes of analysis of stress and strain in cutting, many details must be ignored and a much simplified model must be assumed, even to deal with such an uncomplicated operation as lathe turning.   The making of these simplifications is justified in order to build up a valuable framework of theory, provided it is borne in mind that real-life behaviour can be completely accounted for only if the complexities, which were ignored in the first analysis, are reintroduced.

An important simplification is to ignore both the irregular cross section of real chips and the chip spread, and to assume a rectangular cross section, whose width is the original depth of cut, and whose height is the measured mean thickness of the chip.   With these assumptions, the formation of chips is considered in terms of the simplified diagram, *Figure 3.1*, an idealised section normal to the cutting edge of a tool used in orthogonal cutting.

CHIP FORMATION

In practical tests, the mean chip thickness can be obtained by measuring the length, l, and weight, W, of a piece of chip.   The mean chip thickness t_2 is then

$$t_2 = \frac{W}{\rho w l} \tag{3.1}$$

where ρ = density of work material (assumed unchanged during chip formation)
w = width of chip (depth of cut)

The mean chip thickness is a most important parameter.   In practice the chip is never thinner than the feed, which in orthogonal cutting, is equal to the undeformed chip thickness, t_1 (*Figure 3.1*).   Chip thickness is not constrained by the tooling, and, with many ductile metals, the chip may be five times as thick as the feed, or even more.   The chip thickness ratio t_2/t_1 is geometrically related to the tool rake angle and the *shear plane angle* ϕ (*Figure 3.1*).   The latter is the angle formed between the direction of movement of the workpiece *OA* (*Figure 3.1*) and the *shear plane* represented by the line *OD*, from the tool edge to the position where the chip leaves the work surface.   For purposes of simple analysis the chip is assumed to form by shear along the *shear plane*.   In fact the shearing action takes place in a zone close to this plane.   The shear plane angle is determined from experimental values of t_1 and t_2 using the relationship

$$\cot \phi = \frac{\dfrac{t_2}{t_1} - \sin \alpha}{\cos \alpha} \tag{3.2}$$

and, where the rake angle, α, is $0°$

$$\cot \phi = \frac{t_2}{t_1} \tag{3.3}$$

The chip moves away with a velocity V_c which is related to the cutting speed V and the chip thickness ratio

$$V_c = V \frac{t_1}{t_2} \tag{3.4}$$

If the chip thickness ratio is high, the shear plane angle is small and the chip moves away slowly, while a large shear plane angle means a thin, high-velocity chip.

As any volume of metal, e.g. *klmn* (*Figure 3.1*) passes through the shear zone, it is plastically deformed to a new shape — *pqrs*. The amount of plastic deformation (shear strain, γ) has been shown to be related to the shear plane angle ϕ and the rake angle α by the equation[2]

$$\gamma = \frac{\cos \alpha}{\sin \phi \cos (\phi - \alpha)} \tag{3.5}$$

The meaning of 'shear strain', and of the units in which it is measured, is shown in the inset diagram on the graph *Figure 3.4*. A unit displacement of one face of a unit cube is a shear strain of 1 ($\gamma = 1$). *Figure 3.4* is a graph showing the relationship between the shear strain in cutting and the shear plane angle for three values of the rake angle. For any rake angle there is a minimum strain when the mean chip thickness is equal to the feed ($t_2 = t_1$). For zero rake angle, this occurs at $\phi = 45°$. The change of shape of a unit cube after passing through the shear plane for different values of the shear plane angle is shown in the lower diagram of *Figure 3.4* for a tool with a zero rake angle. The minimum strain at $\phi = 45°$ is apparent from the shape change.

At zero rake angle the minimum shear strain is 2. The minimum strain becomes less as the rake angle is increased, and if the rake angle could be made very large, strain in chip formation could become very small. In practice the optimum rake angle is determined by experience, too large an angle weakening the tool and leading to fracture. Rake angles higher than $30°$ are seldom used and, in recent years, the tendency has been to decrease the rake angle to make the tools more robust, to enable harder but less tough tool materials to be used. (Chapter 6). Thus, even under the best cutting conditions, chip formation involves very severe plastic deformation, resulting in considerable work-hardening and structural change. It is not surprising that metals and alloys lacking in ductility are periodically fractured on the shear plane.

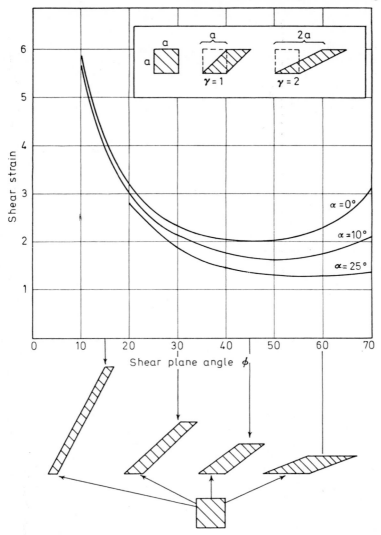

Figure 3.4   Strain on shear plane (γ) vs shear plane angle (φ) for three values of rake angle (α)

The chip/tool interface

The formation of the chip by shearing action at the shear plane is the aspect of metal cutting which has attracted most attention from those who have attempted analyses of machining. Of at least equal importance for the understanding of machinability and the performance of cutting tools is the movement of the chip and of the work material across the faces and around the edge of the tool. In most analyses this has been treated as a classical friction situation, in which 'frictional forces' tend to restrain movement across the tool surface, and the forces have been considered in terms of a coefficient of friction (μ) between the tool and work materials. However, detailed studies of the tool/work interface have shown that this approach is inappropriate to most metal cutting conditions. It is necessary, at this stage, to explain why classical friction concepts do not apply and to suggest a more suitable model for analysing this situation.

The concept of *coefficient of friction* derives from the work of Amonton and Coulomb who demonstrated that, in many common examples of the sliding of one solid surface over another, the force (F) required to initiate or continue sliding is proportional to the force (N) normal to the interface at which sliding is taking place

$$\mu = \frac{F}{N} \tag{3.6}$$

This coefficient of friction (μ) is dependent only on these forces and is independent of the sliding area of the two surfaces. The work of Bowden and Tabor,[3] Archard and many others, has demonstrated that this proportionality results from the fact that real solid surfaces are never completely flat on a molecular scale, and therefore make contact only at the 'tops of the hills', while the 'valleys' are separated by a gap.

The frictional force is that required to separate or shear apart the areas in actual contact at the tops of the hills. Under normal loading conditions for sliding, this real contact area is very small, often less than one thousandth of the apparent area of contact of the sliding surfaces. When the force acting normal to the surface is increased, the area of contact at the tops of the hills is increased in proportion to the load. The frictional force required to shear the contact areas, therefore, also increases proportionately, so that the sliding force is directly proportional to the normal force. Thus the concepts of sliding, friction, and a coefficient of friction are adequate for many engineering situations where stresses between surfaces are small compared with the yield stress of the materials.

When the normal force is increased to such an extent that the real area of contact is a large proportion of the apparent contact area, it is no longer possible for the real contact area to increase proportionately to the load. In the extreme case, where the two surfaces are completely in contact over the whole area, the real area of contact becomes independent of the normal force, and the force required to move one surface over the other becomes that necessary to shear the weaker of the two materials across the whole area. This force is almost independent of the normal force, but is directly proportional to apparent area of contact – a relationship directly opposed to that of classical friction concepts.

It is, therefore, important to know what conditions exist at the interface between tool and work material during cutting. This is a very difficult region to investigate. Few significant observations can be made while cutting is in progress, and the conditions existing must be inferred from studies of the interface after cutting has stopped, and from measurements of stress and temperature. The conclusions presented here are deduced from studies, mainly by optical and electron microscopy, of the interface between work-material and tool after use in a wide variety of cutting conditions. Evidence comes from worn tools, from quick-stop sections and from chips.

The most important conclusion from the observations is that contact between tool and work surfaces is so nearly complete over a large part of the total area of the interface, that sliding at the interface is impossible under most cutting conditions.[4] The evidence for this statement is now reviewed.

When cutting is stopped by disengaging the feed and withdrawing the tool, layers of the work material are commonly, but not always, observed on the worn tool surfaces, and micro-sections through these surfaces can preserve details of the interface. Special metallographic techniques are essential to prevent rounding of the edge where the tool is in contact with a much softer, thin layer of work-material.

Figure 3.5 is a photomicrograph (X 1 500) of a section through the cutting edge of a cemented carbide tool used to cut steel. The white area is the residual steel layer, which is attached to the cutting edge (slightly rounded), the tool rake face (horizontal) and down the worn flank. The two surfaces remained firmly attached during the grinding and polishing of the metallographic section. Any gap larger than 0.1 μm (4 micro-inches) would be visible at the magnification in this photomicrograph, but no such gap can be seen. It is unlikely that a gap exists since none of the lubricant used in polishing oozed out afterwards.

Figure 3.6 is an electron micrograph of a replica of a section through a high speed steel tool at the rake surface, where the work material (top) was adherent to the tool. The tool had been used for cutting a very low carbon steel at high cutting speed (200 m min^{-1}, 600 ft/min). The

Figure 3.5 Section through cutting edge of cemented carbide tool after cutting steel at 84 m min^{-1} (275 ft/min)[4]

22

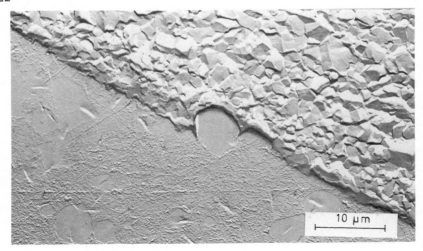

Figure 3.6   Section through rake face of steel tool and adhering metal after cutting iron at high speed.   Etched in Nital.   Electron micrograph of replica

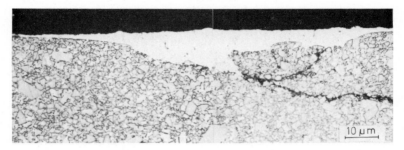

Figure 3.7   Section through rake face of cemented carbide tool, after cutting nickel-based alloys, with adhering work material[4]

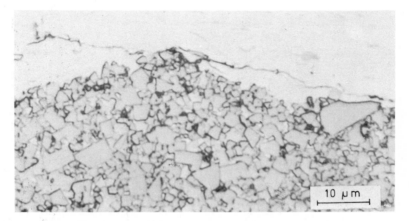

Figure 3.8   Section parallel to rake face of cemented carbide tool, through the cutting edge, after cutting cast iron.   White material is adhering cast iron[4]

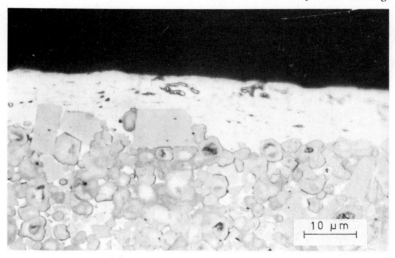

Figure 3.9 As Figure 3.8, *the tool being a steel cutting grade of carbide with two carbide phases*[4]

steel adhering to the tool had recrystallised and contact between the two surfaces is continuous, in spite of the uneven surface of the tool, which would make sliding impossible.

Many investigations have shown the two surfaces to be interlocked, the adhering metal penetrating both major and minor irregularities in the tool surface. *Figure 3.7* shows a nickel based alloy penetrating deeply into a crack on the rake face of a carbide tool. When cutting grey cast iron, much less adhesion might be expected than when the work material is steel or a nickel based alloy, because of the lower cutting forces, the segmented chips and the presence of graphite in the grey iron. However, micro-sections demonstrate a similar extensive condition of seizure. *Figure 3.8* shows a section, parallel to the rake face through the worn flank of a carbide tool used to cut a flake graphite iron at 30 m min⁻¹ (100 ft/min). There is a crack through the adhering work material, which may have formed during preparation of the polished section. If there had been gaps at the interface, the crack would have formed there, and the fact that it did not do so is evidence for the continuity and strength of the bond between tool and work material. *Figure 3.9* is a similar section through the worn edge of a carbide tool containing titanium carbide, as well as tungsten carbide and cobalt. Close contact can be observed between the adhering metal and all the grains of both carbide phases present in the structure of this tool.

The evidence of optical and electron microscopy demonstrates that the surfaces investigated are interlocked or 'seized' to such a degree that sliding, as normally conceived between surfaces with only the high spots in contact, is not possible. Some degree of metallurgical bonding is suggested by the frequently observed persistence of contact through all the stages of grinding, lapping and polishing of sections. There is, however, a considerable variation in the strength of bond generated, depending on the tool and work materials and the conditions of cutting. In some cases grinding and lapping causes the work material to break away with the separation occurring along the interface or part of the interface. In other cases, the whole chip adheres to the tool, even during a quick-stop while the tool is explosively propelled from the cutting position, *Figure 3.10*. A quick-stop may also result in a thin layer of work material staying on the tool. *Figure 3.11* shows such a layer on a tool used to cut a low carbon steel. In this case the bond was so strong that separation during quick-stop took place by a ductile tensile fracture *within the chip*, close to the tool surface. This tensile fracture is seen in *Figure 3.12* which shows part of the layer surface in *Figure 3.11* at high magnification.

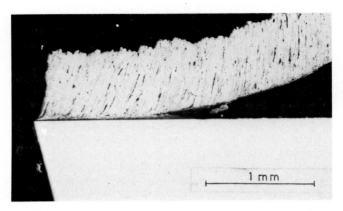

Figure 3.10.   Section through high speed steel tool, with adhering very low carbon steel chip, after quick-stop

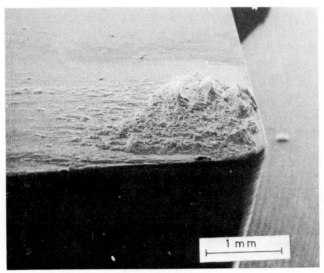

Figure 3.11   Scanning electron micrograph showing adhering metal (steel) on rake face of tool after quick-stop

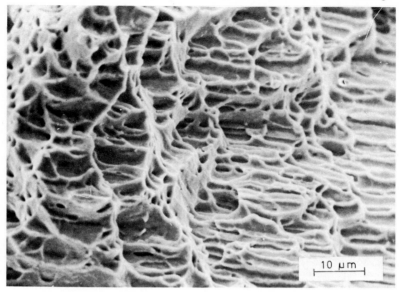

Figure 3.12   Fracture surface on adhering metal in Figure 3.11

It is evidence of this character that has demonstrated the mechanically-interlocked and/or metallurgically-bonded character of the tool-work interface as a normal feature of metal cutting. Under these conditions the movement of the work material over the tool surface cannot be adequately described using the terms 'sliding' and 'friction' as these are commonly understood.   Coefficient of friction is not an appropriate concept for dealing with the relationship between forces in metal cutting for two reasons – (1) there can be no simple relationship between the forces normal to and parallel to the tool surface and (2) the force parallel to the tool surface is not independent of the area of contact, but on the contrary, the area of contact between tool and work material is a very important parameter in metal cutting. The condition where the two surfaces are interlocked or bonded is referred to here as *conditions of seizure* as opposed to *conditions of sliding* at the interface.

The generalisation concerning seizure at the tool/work interface having been stated must now be qualified. The enormous variety of cutting conditions encountered in industrial practice has been discussed and there are some situations where there is sliding contact at the tool surface. It has been demonstrated at very low cutting speed (a few centimetres per minute) and at these speeds sliding is promoted by the use of active lubricants. Sliding at the interface occurs, for example, near the centre of a drill where the action of cutting becomes more of a forming operation. Even under seizure conditions, it must be rare for the *whole* of the area of contact between tool and work surfaces to be seized

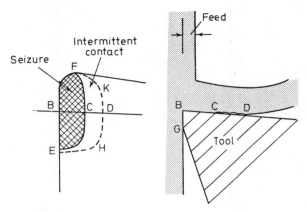

Figure 3.13   Areas of seizure on cutting tool

together. This is illustrated diagrammatically in *Figure 3.13* for a lathe cutting tool. Examination of used tools provides positive evidence of seizure on the tool rake face close to the cutting edge, *BECF* in *Figure 3.13,* the width *BC* being considerably greater than the feed. Beyond the edge of this area there is frequently a region where visual evidence suggests that contact is intermittent, *EHDKFC* in *Figure 3.13*. On the worn flank surface, *BG*, it is uncertain to what extent seizure is continuous and complete, but *Figures 3.5, 3.8* and *3.9* show the sort of evidence demonstrating that seizure occurs on the flank surface also, particularly close to the edge.

Chip flow under conditions of seizure

Since most published work on machining is based implicitly or explicitly on the classical friction model of conditions at the tool/work interface, the evidence for seizure requires reconsideration of almost all aspects of metal cutting theory. We are accustomed to thinking of seizure as a condition in which relative movement ceases, as when a bearing or a piston in a cylinder is seized. Movement stops because there is insufficient force available to shear the metal at the seized junctions, and if greater force is applied, the normal result is massive fracture at some other part of the system. With metal cutting, however, it is necessary to accustom oneself to the concepts of a system in which relative movement continues under conditions of seizure. This is possible because the area of seizure is small, and sufficient force is applied to shear the work material near the seized interface. Tool materials have high yield stress to avoid destruction under the very severe stresses which seizure conditions impose.

Under sliding conditions, relative movement can be considered to take place at a surface which is the interface between the two bodies. Movement occurs at the interface because the force required to shear the contact points is much smaller than that required to shear either of the two bodies. Under seizure conditions it can no longer be assumed that relative movement takes place at the interface, because the force required to overcome the interlocking and bonding is normally higher than that required to shear the adjacent metal. Relative motion under seizure involves bulk shearing in the weaker of the two bodies, in a region of finite thickness which may lie immediately adjacent to the interface or at some distance from it, depending on the stress system involved.

In sections through chips and in quick-stop sections, zones of intense shear near the interface are normally observed, except under conditions where sliding takes place. *Figure 3.14* shows the sheared zone adjacent to the tool surface in the case of steel being cut at high speed. The thickness of these zones is often of the order of $25-50\,\mu$ m (0.001–0.002 in), and strain within the regions is much more severe than on the shear plane, so that normal structural features of the metal or alloy being cut are greatly altered or completely transformed. The behaviour of the work material in these regions is, in many ways, more like that of an extremely viscous liquid than that of a normal solid metal. For this reason, the term *flow-zone* is used to describe them. As can be seen from *Figure 3.14,* there is not a sharp line separating the flow-zone from the body of the chip, but a gradual blending in. There is in fact a *pattern of flow* in the work material around the cutting edge and across the tool faces, which is characteristic of the metal or alloy being cut and the conditions of cutting. A pattern of flow and a velocity gradient within the work material, approaching zero at the tool/work interface, are the basis of the model for relative movement under conditions of seizure, to replace the classical friction model of sliding conditions.

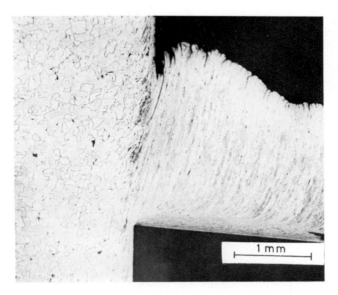

Figure 3.14 Section through quick-stop showing flow-zone in very low carbon steel after cutting at high speed

The built-up edge

It is the condition of seizure which gives rise to one of the major types of chip formation.   When cutting many alloys under suitable conditions, hardened work material, adhering around the cutting edge and along the rake face, accumulates to displace the chip from immediate direct contact with the tool as shown in *Figure 3.15*.   The *built-up edge,* which occurs frequently under industrial cutting conditions, can be formed with either a continuous or a discontinuous chip – for example when cutting steel or when cutting cast iron.   The built-up edge is discussed later in more detail, but a number of its features are basic to an understanding of machining in general.

The built-up edge is not a separate body of metal during the cutting operation.   Diagrammatically it should be depicted as in *Figure 3.16*. The new work surface is being formed at *A* and the under surface of the chip at *B,* but between *A* and *B* the built-up edge and the work material are one continuous body of metal, not separated by free surfaces. In effect the flow-zone has been transferred from the tool surface to the top of the built-up edge.   This illustrates the principle that, under seizure conditions, relative movement does not necessarily take place immediately adjacent to the interface.

The built-up edge is a dynamic structure, being constructed of successive layers greatly hardened under extreme strain conditions.   The size of the built-up edge cannot increase indefinitely, the resolved shear stresses

0.5 mm

Figure 3.15   Section through quick-stop showing built-up edge in steel

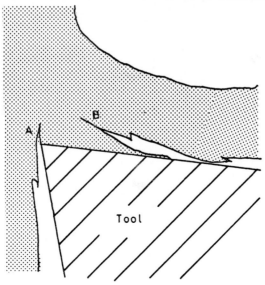

Figure 3.16   Form of built-up edge[4]

increasing until part of the structure is sheared off and carried away on
the work surface or on the underside of the chip as seen in *Figure 3.15*.
In some cases the resolved stress on the built-up edge becomes such as to
cause the whole build-up to be sheared from the tool surface, followed by
the formation of a new build-up.

There is no hard and fast line between a built-up edge (*Figure 3.15*)
and a flow-zone (*Figure 3.14*). Seizure between tool and work material
is a feature of both situations and every shade of transitional form
between the two can be observed. The built-up edge occurs in many
shapes and sizes and it is not always possible to be certain whether or
not it is present.

Machined surfaces

Figures 3.14 and *3.15* demonstrate that machined surfaces are formed by
a process of fracture under shearing stresses. With ductile metals and
alloys, both sides of a shear fracture are plastically strained, so that
some degree of plastic strain is a feature of machined surfaces. The
amount of strain and the depth below the machined surface to which it
extends, can vary greatly, depending on the material being cut, the tool
geometry, and the cutting conditions, including the presence or absence
of a lubricant.[5] The deformed layer on the machined surface can be
thought of as that part of the flow-pattern around the cutting edge which
passes off with the work material, so that an understanding of the flow-
pattern, and the factors which control it, is important in relation to the

character of the machined surface.   The presence or absence of seizure on those parts of the tool surface where the new work surface is generated can have a most important influence, as can the presence or absence of a built-up edge and the use of sharp or worn tools.   These factors influence not only the plastic deformation, hardness and properties of the machined surface, but also its roughness, its precise configuration and its appearance.

References

1.   WILLIAMS, J.E., SMART, E.F. and MILNER, D.R., *Metallurgia,* **81**, 6 (1970)
2.   HILL, R., *Plasticity,* Oxford – Clarendon Press (1950)
3.   BOWDEN, F.P. and TABOR, D., *Friction and Lubrication of Solids,* Oxford University Press (1954)
4.   TRENT, E.M., *I.S.I. Special Report,* **94**, 11 (1967)
5.   CAMATINI, E. *Proc. 8th Int. Conf. M.T.D.R., Manchester* (1967)

4

FORCES IN METAL CUTTING

The forces acting on the tool are an important aspect of machining. For those concerned with the manufacture of machine tools, a knowledge of the forces is needed for estimation of power requirements and for design of structures adequately rigid and free from vibration. The cutting forces vary with the tool angles, and accurate measurement of forces is helpful in optimising tool design. Scientific analysis of metal cutting also requires knowledge of the forces, and in the last eighty years many dynamometers have been developed, capable of measuring tool forces with considerable accuracy.

Detailed accounts of the construction of dynamometers for this work are available in the technical literature.[1,2] All methods used are based on measurement of elastic deflection of the tool under load. For a semi-orthogonal cutting operation in lathe turning, the force components can be measured in three directions, *Figure 4.1*, and the force relationships are relatively simple. The component of the force acting on the rake face of the tool, normal to the cutting edge, in the direction *YO* is called here the *cutting force, F_c*. This is usually the largest of the three force components, and acts in the direction of the cutting velocity. The

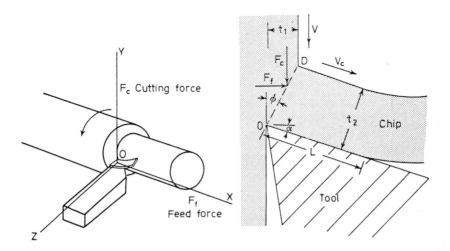

Figure 4.1   Forces acting on cutting tool

force component acting on the tool in the direction *OX,* parallel with the direction of feed, is referred to as the *feed force, F_f.* The third component, acting in the direction *OZ,* tending to push the tool away from the work in a radial direction, is the smallest of the force components in semi-orthogonal cutting and, for purposes of analysis of cutting forces in simple turning, it is usually ignored and not even measured.

The deflection of the tool under load was measured by dial gauges in the early dynamometers. Being relatively insensitive, they could measure only rather large movements and the tools had to be reduced in section to increase the deflection under load. Reduced rigidity of the tool restricted the range of cutting conditions which could be investigated. The performance of tool dynamometers has been greatly improved by the introduction of wire strain-gauges, transducers or piezo-electric crystals as sensors to measure deflection. More rigid tools can now be used and fluctuations of forces over very small time intervals can be recorded. The forces involved in machining are relatively low compared with those in other metal working operations such as forging. Because the layer of metal being removed – the chip – is thin, the forces to be measured are usually not greater than a few tens or hundreds of kilograms.

Stress on the shear plane

Measurement of the forces and the chip thickness make it possible to explore the stresses for simple, orthogonal cutting conditions, as described in Chapter 3. For conditions where a continuous chip is formed with no built-up edge, the work is sheared in a zone close to the shear plane (*OD* in *Figure 4.1*) and, for the purposes of this simple analysis, it is assumed that shear takes place *on* this plane to form the chip. The force acting on the shear plane, F_s, is calculated from the measured forces and the shear plane angle:

$$F_s = F_c \cos \phi - F_f \sin \phi \qquad (4.1)$$

The shear stress required to form the chip, k_s, is

$$k_s = \frac{F_s}{A_s} \qquad (4.2)$$

where A_s = area of shear plane.

The force required to form the chip is dependent on the shear yield strength of the work material under cutting conditions, and on the area of the shear plane. Many calculations of shear yield strength in cutting have been made, using data from dynamometers, and chip thickness measurements. In general the shear strength of metals and alloys in cutting has been found to vary only slightly over a wide range of cutting speeds and feed rates, and the values obtained are not greatly different from the yield strengths of the same materials measured in laboratory tests at appropriate amounts of strain. *Table 4.1* shows the values of k_s measured during cutting, for a variety of metals and alloys.

Table 4.1

Material	*Shear yield strength in cutting* k_s	
	ton f/in^2	N mm^{-2}
Iron	24	370
0.13% C steel	31	480
Ni-Cr-V steel	45	690
Austenitic stainless steel	41	630
Nickel	27	420
Copper (annealed)	16	250
Copper (cold worked)	17	270
Brass (70/30)	24	370
Aluminium	6.3	97
Magnesium	8	125
Lead	2.3	36

Provided the shear plane area remains constant, the force required to form the chip, being dependent on the shear yield strength of the metal, is increased by any alloying or heat treatment which raises the yield strength. In practice, however, the area of the shear plane is very variable, and it is this area which exerts the dominant influence on the cutting force, often more than outweighing the effect of the shear strength of the metal being cut.

In orthogonal cutting the area of the shear plane is geometrically related to the undeformed chip thickness t_1 (the feed), to the chip width w (depth of cut) and to the shear plane angle ϕ.

$$A_s = \frac{t_1 \, w}{\sin \phi} \tag{4.3}$$

The forces increase in direct proportion to increments in the feed and depth of cut, which are two of the major variables under the control of the machine tool operator. The shear plane angle, however, is not directly under the control of the machinist, and in practice it is found to vary greatly under different conditions of cutting, from a maximum of approximately 45° to a minimum which may be 5° or even less. *Table 4.2* shows how the chip thickness, t_2, the area of the shear plane, and the shearing force, F_s, for a low carbon steel, vary with the shear plane angle for orthogonal cutting at a feed rate of 0.5 mm/rev and a depth of cut of 4 mm.

Table 4.2

Shear plane angle ϕ	Chip thickness t_2 mm	Shear plane area A_s mm^2	Shearing force F_s N
45°	0.50	2.8	1 340
35°	0.71	3.5	1 680
25°	1.07	4.7	2 260
15°	1.85	7.7	3 700
5°	5.75	23.0	11 000

Thus, when the shear plane angle is very small, the *shearing force* may be more than five times that at the minimum where $\phi = 45°$, under conditions where the *shear stress* of the work material remains constant. It is important, therefore, to investigate the factors which regulate the shear plane angle, if cutting forces are to be controlled or even predicted. Much of the work done on analysis of machining has been devoted to methods of predicting the shear plane angle.

Forces in the flow-zone

Before dealing with the factors determining the shear plane angle, consideration must be given to the other main region in which the forces arise – the rake face of the tool. For the simple case where the rake angle is 0°, the feed force F_f is a measure of the drag which the chip exerts as it flows away from the cutting edge across the rake face. The origin of this resistance to chip flow is discussed in Chapter 3, where the normal existence of conditions of seizure over a large part of the interface between the under surface of the chip and the rake face of the tool is demonstrated. Although there are areas where sliding occurs, at the periphery of the seized contact region, the force to cause the chip to move over the tool surface is mainly that required to shear the work material in the flow-zone across the area of seizure. Under most cutting conditions, the contribution to the feed force made by friction in the non-seized areas is probably relatively small. The feed force, F_f, can, therefore, be considered as the product of the shear strength of the work material at this surface (k_r) and the area of seized contact on the rake face (A_r).

$$F_f = A_r k_r \tag{4.4}$$

While the feed force can be measured accurately, the same cannot be said for the area of contact, which is usually ill-defined. Observation of this area during cutting is not possible, and when the tools are examined after use, worn areas, deposits of work material as smears and small lumps, and discolouration due to oxidation or carbonisation of cutting

oils are usually seen.   The deposits may be on the worn areas, or in
adjacent regions and the limits of the worn areas are often difficult to
observe.   Many of these effects are relatively slight and not easily visible
on a ground tool surface.   In experimental work on these problems, therefore, the use of tools with polished rake faces is strongly recommended,
unless the quality of the tool surface is a parameter being investigated.
There is no universal method for arriving at an estimate of the contact
area, the problem being different with different work materials, tool
materials and cutting conditions.   It is necessary to adopt a critical
attitude to the criteria defining the area of contact, and a variety of
techniques may have to be employed in a detailed study of the used
tool.

It may be possible to dissolve chemically the adhering work material
to expose the worn tool surface for examination.   This technique is used
with carbide tools after cutting steel or iron, but it cannot usually be
applied with steel tools because reagents effective in removing work material also attack the tool.   A quick-stop examination of sections through
the tool surfaces may be required.   An example in *Figure 4.2* shows the
rake surface of a high speed steel tool used to cut titanium.   A quick-
stop technique had been used and the area of complete seizure is, in
this case, defined by the demarcation line *AB* which encloses an area
within which the adhering titanium shows a tensile fracture, where it had

Figure 4.2   Contact area on rake face of tool used to cut titanium

separated through the under surface of the chip. Beyond this line small smears of titanium indicate a region where contact had been occasional and short lived.

In orthogonal cutting the width of the contact region is usually equal to the depth of cut, or only slightly greater, although with very soft metals there may be a more considerable chip spread. The length of contact (*L* in *Figure 4.1*) is always greater than the undeformed chip thickness t_1, and may be as much as ten times longer; it is usually uneven along the chip width, and a mean value must be estimated. The contact area is mainly controlled by the length of contact *L*. It is a most important parameter, having a very large influence on cutting forces, on tool life and on many aspects of machinability.

Early analyses of forces in cutting considered the movement of the chip across the tool in terms of classical friction theory, in which the important relationship was the coefficient of friction, measured by the ratio of frictional to normal force on the rake face of the tool.[3,4]   With this model, area of contact was not significant, no attempt was made to measure it, and much valuable information was lost. When an estimate can be made of the area of contact, A_r, the mean shear strength of the work material at the rake face k_r, can be calculated, equation 4.4. It is doubtful whether the values of k_r are of much significance, partly because of the inaccuracies in measurement of contact length, but also because of the extreme conditions of shear strain, strain rate, temperature and temperature gradient which exist in this region. The values of k_r are not likely to be the same as those of the shear stress on the shear plane, k_s, and are usually lower, decreasing with increasing cutting speed.

The feed force F_f is increased by any changes in composition or structure of the work material which increase k_r. Alloying may either increase or reduce k_r and this is discussed in Chapter 7.

The shear plane angle and minimum energy theory

We can now return to the problem of the shear plane angle. The thickness of the chip is not constrained by the tool, and the question is — what *does* determine whether there is a thick chip with a small shear plane angle and high cutting force, or a thin chip with large shear plane angle and minimum cutting force? Consideration of the energy expended in cutting provides at least a qualitative answer to this question. The explanation given here is a simplified version of the treatment of the subject proposed by G.W. Rowe and P.T. Spick.[5]

The explanation is based on the hypothesis that, since it is not externally constrained, the shear plane will adopt such a position that the total energy expended in the system (energy on the shear plane plus energy on the rake face) is a minimum.

Consider first the rate of work done on the shear plane. It can be shown that

$$\frac{dW_s}{dt} = F_s \, V_s \tag{4.5}$$

where W_s = work done on shear plane
V_s = rate of strain on shear plane.
But

$$V_s = \frac{V}{\cos \phi} \tag{4.6}$$

where V = cutting speed.

For the simple conditions where the rake angle $\alpha = 0°$, it has been shown that

$$F_s = \frac{t_1 \, w \, k_s}{\sin \phi} \tag{4.7}$$

$$\therefore \quad \frac{dW_s}{dt} = \frac{t_1 \, w \, k_s \, V}{\sin \phi \, \cos \phi} \tag{4.8}$$

Thus if all the cutting conditions are held constant, it is found that the rate of work on the shear plane is proportional to $1/\sin \phi \cos \phi$, which has been shown to be the amount of strain (γ) on the shear plane (equation 3.5) where the rake angle (α) = 0°. Curve *1* in *Figure 4.3* shows how the work done on the shear plane varies with the shear plane angle, and this has, of course, the same shape as the curve for zero rake angle in *Figure 3.4,* with a minimum at $\phi = 45°$. Thus, if the feed force F_f and the work done on the rake face are so small that they can be neglected, the minimum energy theory proposes that the shear plane angle would be 45°, with the chip thickness t_2 equal to the feed t_1. Where the rake angle is higher than zero, the minimum work is at a shear plane angle higher than 45°, but always at a value where $t_2 = t_1$. In practice, the chip is sometimes approximately equal in thickness to the feed, never thinner, but often much thicker.

To assess the total energy consumed, the work done on the rake face must also be considered. Again, for the simple conditions where $\alpha = 0°$:

$$\frac{dW_r}{dt} = F_f \, V_c \tag{4.9}$$

W_r = work done at the rake face

where
V_c = velocity of chip
$V_c = V \tan \phi \tag{4.10}$

and, where there is no chip spread,

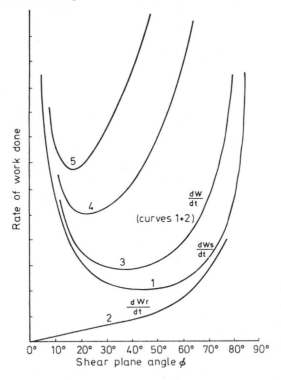

Figure 4.3   Rate of work done vs shear plane angle φ.   (After Rowe and Spick[5])

From equation 4.4 $F_f = w L k_r$ (4.11)

$$\frac{dW_r}{dt} = w L k_r V \tan \phi$$ (4.12)

Thus, if all cutting conditions are held constant, including L, the length of contact on the rake face, the rate of doing work on the rake face is proportional to tan φ.   From curve *2*, *Figure 4.3*, dW_r/dt is seen to increase with the shear plane angle and becomes very large as φ approaches 90°.

The total rate of work done, dW/dt is represented by curve *3*, which is a sum of curves *1* and *2*, *Figure 4.3*.   In curve *3* the minimum work done occurs at a value of φ less than 45° and at an increased rate of work.   Curve *2* is plotted for an arbitrary, small value of L and of k_r. The contact length L is one of the major variables in cutting, and its influence can be demonstrated by plotting a series of curves for different values of L.   Curves *4* and *5* show the total rate of work done for values of L four and eight times that of curve *3*.   This family of curves shows that, as the contact length on the rake face of the tool increases, the minimum energy occurs at lower values of the shear plane angle and the rate of work done increases greatly. A similar reduction in ϕ would result from increases in the value of the yield stress, k_r.

It has already been explained that, at low values of ϕ, the chip is thick, the area of the shear plane becomes larger, and, therefore the cutting force, F_c, becomes greater. **Thus the consequence of increasing either the shear yield strength at the rake face, or the contact area (length) is to raise not only the feed force F_f but also the cutting force F_C. The contact area on the tool rake face in particular is seen to be a most important region, controlling the mechanics of cutting, and becomes a point of focus for research on machining.** Not only the forces, but temperatures, tool wear rates, and the machinability of work materials are closely associated with what happens in this region which receives much attention in the rest of this book.

Forces in cutting metals and alloys

Cutting forces have been measured in research programmes on many metals and alloys, and some of the major trends are now considered.[6,7] When cutting many metals of commercial purity the forces are found to be high; this is true of iron, nickel, copper and aluminium, among others. With these metals, the area of contact on the rake face is found to be very large, the shear plane angle is small and the very thick, strong chips move away at slow speed. For these reasons, pure metals are notoriously difficult to machine.

The large contact area is probably associated with the high ductility of these pure metals, but the reasons are not completely understood. That the high forces *are* related to the large contact area can be simply demonstrated by cutting these commercially pure metals with specially shaped tools on which the contact area is artificially restricted. This is shown diagrammatically in *Figure 4.4*. In *Table 4.3* an example is given of the

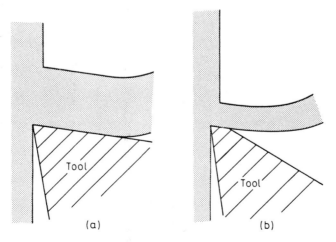

Figure 4.4   (a) Normal tool; (b) tool with restricted contact on rake face

forces, shear plane angle and chip thickness when cutting a very low carbon steel (in fact, a commercially pure iron) at a speed of 91.5 m min^{-1} (300 ft/min) a feed rate of 0.25 mm (0.010 in)/rev feed and depth of cut of 1.25 mm (0.05 in). In the first column are the results for a normal tool, *Figure 4.4a*, and in the second column those for a tool with contact length restricted to 0.56 mm (0.022 in), *Figure 4.4b*.

Table 4.3

	Normal tool	*Tool with restricted contact*
Cutting force F_c	1 400 N 315 lbf	670 N 150 lbf
Feed force F_f	1 310 N 295 lbf	254 N 51 lbf
Shear plane angle ϕ	7° 58′	21° 39′
Chip thickness t_2	1.83 mm 0.072 in	0.66 mm 0.026 in

Reduction in forces by restriction of contact on the rake face may be a useful technique in some conditions, but in many cases it is not practical because it weakens the tool.

Not all pure metals form such large contact areas, with high forces. For example, when cutting commercially pure magnesium, titanium and zirconium the forces and contact areas are much smaller and the chips are thin.

It is common experience, when cutting most metals and alloys, that the chip becomes thinner and forces *decrease* as the cutting speed is raised. *Figure 4.5* shows force/cutting speed curves for iron, copper and titanium at a feed rate of 0.25 mm rev^{-1} (0.010 in/rev) and a depth of cut of 1.25 mm (0.050 in). The decrease in both F_c and F_f with cutting speed is most marked in the low speed range. This drop in forces is partly caused by a decrease in contact area and partly by a drop in shear strength (k_r) in the flow-zone as its temperature rises with increasing speed. This is discussed in Chapter 5.

Alloying of a pure metal normally increases its yield strength, but often reduces the tool forces, because the contact length on the rake face becomes shorter. For example, *Figure 4.6* shows force/cutting speed curves for iron and a medium carbon steel, for copper and a 70/30 brass.[6] In each case the tool forces are lower for the alloy than for the pure metal over the whole speed range, the difference being greatest at low speeds. The kink in the curve for the carbon steel in the medium speed range illustrates the effect of a built-up edge, a phenomenon observed when cutting two phased alloys but not pure metals.[8] With steels, a built-up edge forms at fairly low speeds and disappears when the speed is raised. Where it is present the forces are usually abnormally low because the built-up edge acts like a restricted contact tool, effectively reducing contact on the rake face (*Figure 3.15*).

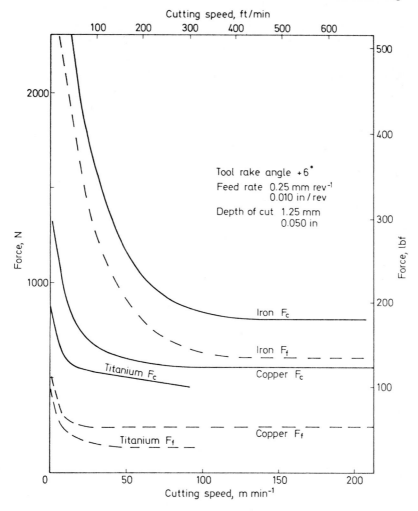

Figure 4.5 Cutting force vs *cutting speed for iron, titanium and copper. (From data by Williams, Smart and Milner[6])*

The tool forces are influenced also by tool geometry, the most important parameter being the rake angle.   Increase in the rake angle lowers both cutting force and feed force, *Table 4.4,* but reduces the strength of the tool edge and may lead to fracture.   The strongest tool edge is achieved with negative rake angle tools, and these are frequently used for the harder grades of carbide and for ceramic tools which lack toughness.   The high forces make negative rake angle tools unsuitable for machining slender shapes which may be deflected or distorted by the high stresses imposed on them.   Tool forces usually rise as the tool is worn, as the clearance angle is destroyed and the area of contact on the clearance face is increased by flank wear.   The forces acting on the tool

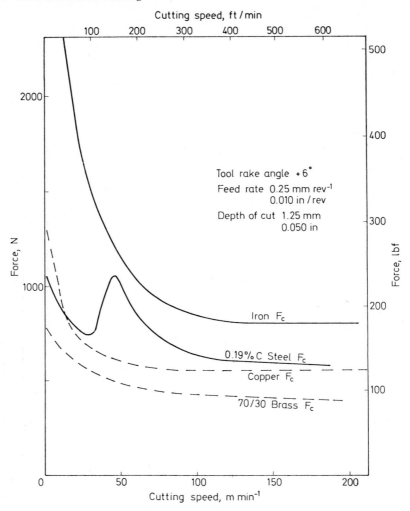

Figure 4.6 Cutting force vs *cutting speed for iron, steel, copper and brass. (From data by Williams, Smart and Milner[6])*

are one of the factors which must be taken into consideration in the design of cutting tools — a very complex and important part of machining technology.

The tool material can also influence the tool forces. When one major type of tool material is substituted for another, the forces may be altered considerably, even if the conditions of cutting and the tool geometry are kept constant. This is probably caused mainly by changes in the area of seized contact.

Finally, the contact length and tool forces may be greatly influenced by cutting lubricants. When cutting at very low speed the lubricant may act to prevent seizure between tool and work and thus greatly reduce the

Table 4.4

Rake angle α	Cutting force F_c				Feed force F_f			
	Feed 0.10 mm rev⁻¹ (0.004 in/rev)		0.20 mm rev⁻¹ (0.008 in/rev)		Feed 0.10 mm rev⁻¹ (0.004 in/rev)		0.20 mm rev⁻¹ (0.008 in/rev)	
	N	lbf	N	lbf	N	lbf	N	lbf
+5°	913	205	-	-	392	88	-	-
+10°	840	189	1520	342	289	65	520	117
+15°	743	167	1328	298	200	45	320	72
+20°	716	161	1210	272	151	34	222	50
+25°	627	141	1158	260	80	18	116	26
+30°	600	135	1090	245	49	11	45	10

Work material – low carbon free cutting steel. Cutting speed 27 m min⁻¹ (90 ft/min)

forces. In the speed range used in most machine shop operations it is not possible to *prevent* seizure near the edge, but liquid or gaseous lubricants, by penetrating from the periphery, can *restrict* the area of seizure to a small region. The action of lubricants is discussed in Chapter 8, but in relation to forces, it is important to understand that they can act to reduce the seized contact area, and thus the forces acting on the tool. They are most effective in doing this at low cutting speeds and become largely ineffective in the high speed range.[9]

Stresses on the tool

With the complex tool configurations and cutting conditions of industrial machining operations, accurate estimation of the localised stresses acting on the tool near the cutting edge defies the analytical methods available. In fact cutting tools are rarely, if ever, designed on the basis of stress calculations; trial and error methods and accumulated experience form the basis for tool design. However, in trying to understand the properties required of tool materials, it is useful to have some knowledge of the general character of these stresses.

In a simple turning operation, two stresses of major importance act on the tool:

1. The cutting force acting on a tool with a small rake angle imposes a stress on the rake face which is largely compressive in character. The mean value of this stress is determined by dividing the cutting force, F_c, by the contact area. Since the contact area is usually not known accurately, there is considerable error in estimation of the mean compressive stress, but the values can be very high when cutting materials of high strength. Examples of estimated values for the mean compressive stress are shown in *Table 4.5* to give an idea of the stresses involved. Unlike the cutting *force,* the compressive *stress* acting on the tool is related to the shear strength of the work material. High *forces* when cutting a pure metal are an indirect result of the large contact area, and the mean *stress* on the tool is relatively low, compared to that imposed when cutting an alloy of the metal.

Table 4.5

Work material	Cutting force F_c		Contact area		Mean compressive stress	
	N	lbf	mm^2	in^2	N mm^{-2}	tonf/in^2
Iron	1 070	240	3.1	0.004 8	340	22
Copper	4 150	930	13.5	0.021	310	20
Titanium	455	102	0.77	0.001 2	570	37
Steel (medium carbon)	490	110	0.65	0.001	770	50
70/30 brass	500	111.5	12.2	0.019	420	27
Lead	323	72.5	22.5	0.035	14	0.9

2. The feed force F_f imposes a shearing stress on the tool over the area of contact on the rake face. The mean value of this stress is equal to the force F_f divided by the contact area. Since F_f is normally smaller than F_c the shear stress is smaller than the compressive stress acting on the same area. Frequently the mean shear stress is between 30% and 60% of the value of the mean compressive stress.

When a worn surface is generated on the clearance face of a tool ('flank wear') both compressive and shear stresses act on this surface. Although the contact area on the flank is sometimes clearly defined, it is very difficult to arrive at values for the forces acting on it, and there are no reliable estimates for the stress on the worn flank surface.

Other stresses acting on the body of the tool are related to the general construction of the tool and to the rigid connection where the tool is fastened to the machine tool. In a lathe, where the tool acts as a cantilever, there are bending stresses giving tension on the upper surface between the contact area and the tool holder. In a twist drill the stresses are mainly torsional. Tools must be strong enough and rigid enough to resist fracture and to give minimum deflection under load. This is an important area of tool design, but is not further discussed because stress is being considered here in relation to tool life and chip formation.

Stress distribution

The *mean* compressive and shear stresses acting on the tool are of significance, but the stress is not evenly distributed over the contact area and knowledge of stress distribution is essential. Methods have been developed for measuring stress distribution under certain simplified laboratory conditions.[10,11] One of these uses a tool made of a photo-elastic material — a polymer, often PVC. Because of the low strength of the polymer, and the rapid drop in strength with temperature, these tools can be used for the cutting only of soft metals of low melting point such as lead and tin, and the cutting speed must be kept low.

Figure 4.7 is a photograph of a photo-elastic tool taken with monochromatic light, while cutting lead at 3.1 m min⁻¹ (10 ft/min) and a feed rate of 0.46 mm rev⁻¹ (0.0185 in/rev).[11] The dark and light bands are regions of equal strain within the tool and, from these, the distribution of compressive and shear stress can be determined. *Figure 4.8* is an example of the type of stress distribution found.

The distribution of compressive stress has been measured also by using as tools a series of steels of different hardness.[12] If a tool has adequate hardness (yield stress) it is only elastically deformed during cutting, but if it is too soft, the tool edge is permanently deformed downward. An estimation can be made of the distribution of stress by cutting a workpiece with a series of tools of decreasing yield strength, and then measuring the permanent deformation of each tool edge. This method has been

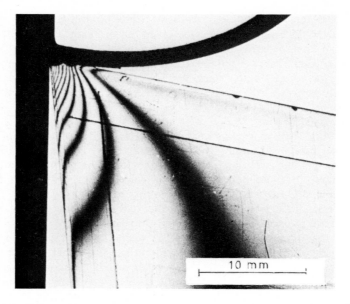

Figure 4.7   Stress distribution in photo-elastic model tool when cutting lead. (Courtesy E. Amini)

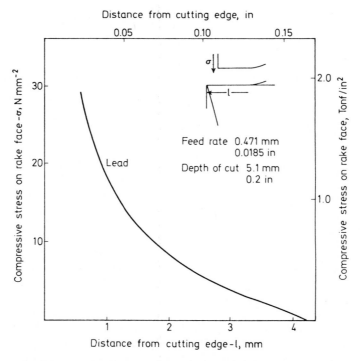

Figure 4.8   Compressive stress on rake face calculated from photo elastic tool as Figure 4.7.   (From data by E. Amini[11])

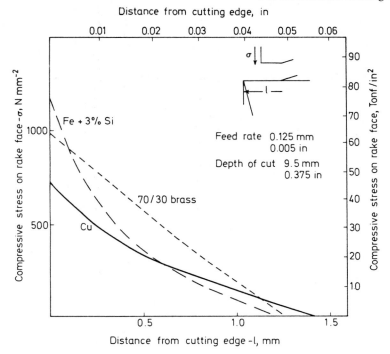

Figure 4.9 *Compressive stress on rake face of tool used to cut copper, brass and iron-silicon alloy. (From data by A.B. Wilcox)*

used to determine stress distribution in tools used to cut copper, aluminium and some of their alloys. Low cutting speeds only are used, to avoid heating of the tool, which would alter its yield stress. *Figure 4.9* shows the results of a series of tests with different metals and alloys.

Although both methods are confined to a narrow range of metals and cutting conditions, the results are useful. They provide evidence that the compressive stress is highest near the cutting edge, diminishing across the rake face to zero where the chip breaks contact with the tool. It seems probable that this stress distribution is common during cutting, except in the presence of a built-up edge, and that the compressive stress at the edge is often double the mean stress, or even greater. The *shear stress* acting on the rake face may be less variable across the contact area, but the distribution of shear stress in this region may be greatly complicated by extremely steep temperature gradients and complex flow patterns.

The very high normal stress levels account for the conditions of seizure on the rake face, particularly near the cutting edge. Comparison can be made with the process of friction welding, in which the joint is made by rotating one surface against another under pressure. With steel, for example, complete welding can be accomplished where the relative speed of the two surfaces is 16–50 m min^{-1} (50–150 ft/min) and the pressure is 45–75 N mm^{-2} (3–5 tonf/in^2). The stress on the rake face in cutting steel may be five or ten times as high as this near the tool edge.   Seizure between the two surfaces is, therefore, the normal condition to be expected under cutting conditions.

The existing knowledge of stress and stress distribution at the tool-work interface is far too scanty to enable tools to be designed on the basis of the localised stresses encountered.   Even for the simplest type of tooling only a few estimates have been made, using a two-dimensional model, where the influence of a tool nose does not have to be considered, and speeds are so low that temperatures are not a problem.   However, the present level of knowledge is useful in relation to analyses of tool wear and failure, the properties required of tool materials and the influence of tool geometry on performance.

References

1.   BOOTHROYD, G. *Fundamentals of Metal Machining.*   Arnold, 115 (1965)
2.   LOEWEN, E.G., MARSHALL, E.R. and SHAW, M.C., *Proc. Soc. Exp. Stress Analysis*, **8**, 1 (1951)
3.   MERCHANT, M.E., *J. Appl. Phys.*, **16**, No.5, 267 (1945)
4.   PALMER, W.B. and OXLEY, P.L.B., *Proc. I. Mech. E.*, **173**, 623 (1959)
5.   ROWE, G.W. and SPICK, P.T.   *Trans. A.S.M.E.*, **89B**, 530 (1967)
6.   WILLIAMS, J.E., SMART, E.F. and MILNER, D.R., *Metallurgia*, **81**, 3, 51, 89 (1970)
7.   EGGLESTON, D.M., HERZOG, R. and THOMSON, E.G., *J. of Engineering for Industry*, August, 263 (1959)
8.   WILLIAMS, J.E. and ROLLASON, E.C., *J. Inst. Met.*, **98**, 144 (1970)
9.   ROWE, G.W. and SMART, E.F., *Proc. 3rd Lubrication Conv., London*, 83, (1965)
10.   ZOREV, N.N., *International Research in Production Engineering, Pittsburgh*, 42 (1963)
11.   AMINI, E., *J. Strain Analysis*, **3**, 206 (1968)
12.   ROWE, G.W. and WILCOX, A.B. *J.I.S.I.*, **209**, 231 (1971)

5

HEAT IN METAL CUTTING

The power consumed in metal cutting is largely converted into heat near the cutting edge of the tool, and many of the economic and technical problems of machining are caused directly or indirectly by this heating action. The cost of machining is very strongly dependent on the rate of metal removal, and may be reduced by increasing the cutting speed and/or the feed rate, but there are limits to the speed and feed above which the life of the tool is shortened excessively. This may not be a major constraint when machining aluminium and magnesium and certain of their alloys, in the cutting of which other problems, such as the ability to handle large quantities of fast moving swarf, may limit the rate of metal removal. The bulk of cutting, however, is carried out on steel and cast iron, and it is in the cutting of these, together with the nickel-based alloys, that the most serious technical and economic problems occur. With these higher melting point metals and alloys, the tools are heated to high temperatures as metal removal rate increases, and above certain critical speeds, the tools tend to collapse after a very short cutting time under the influence of stress and temperature. That heat plays a part in machining was clearly recognised by 1907 when F.W. Taylor in his paper 'On the Art of Cutting Metals' surveyed the steps which had led to the development of the new high speed steels.[1] These, by their ability to cut steel and iron with the tool running at a much higher temperature, raised the permissible rates of metal removal by a factor of four. The limitations imposed by cutting temperatures have been the spur to the tool material development of the last 70 years. Problems remain however, and even with present day tool materials, cutting speeds may be limited to 15 m min^{-1} (50 ft/min) or less when cutting certain creep resistant alloys.

It is, therefore, important to understand the factors which influence the generation of heat, the flow of heat and the temperature distribution in the tool and work material near the tool edge. This was clear to J.T. Nicolson who studied metal cutting in Manchester about the year 1900. In *The Engineer* in 1904[2] he stated 'There is little doubt that when the laws of variation of the temperature of the shaving and tool with different cutting angles, sizes and shapes of cut, and of the rate of abrasion ..... are definitely determined, it will be possible to indicate how a tool should be ground in order to meet with the best efficiency .... the various conditions to be found in practice.' However, determination of temperatures and temperature distribution in the vitally important region near the cutting edge is technically difficult, and progress has been slow in the 70 years since the problem was clearly stated.   Recent research is clarifying some of the principles, but the work done so far is only the beginning of the fundamental survey that is required.

Heat in chip formation

The work done in (1) deforming the bar to form the chip and (2) moving the chip and freshly cut work surface over the tool is nearly all converted into heat.   Because of the very large amount of plastic strain, it is unlikely that more than 1% of the work done is stored as elastic energy, the remaining 99% going to heat the chip, the tool and the work material.

Under most normal cutting conditions, the largest part of the work is done in forming the chip at the shear plane.   In a simplified model of the action on the shear plane, the work material is heated instantaneously as it is sheared and all the heat is carried away by the chip.   Using this model, the temperature of the chip increases according to the equation[3]

$$T_c = \frac{k\gamma}{J\rho c} \tag{5.1}$$

where T_c = temperature increase in chip body
   k = shear flow stress
   γ = shear strain
   J = mechanical equivalent of heat
   ρc = Volume specific heat.

A reasonable estimate of the temperature increase in the body of the chip can be made using this simplified relationship for many conditions of cutting, particularly at relatively high speeds.   The temperature of the chip can affect the performance of the tool only as long as the chip remains in contact; the heat remaining in the chip after it breaks contact is carried out of the system.   Any small element of the *body* of the chip, after being heated in passing through the shear zone, is not further deformed and heated as it passes over the rake face, and the time required to pass over the area of contact is very short.   For example, at a cutting speed of 50 m min^{-1} (150 ft/min), if the chip thickness ratio is 2, the chip velocity is 25 m min^{-1} (75 ft/min).   If the contact on the rake face (L) is 1 mm long, a small element of the chip will pass over this area in just over 2 milliseconds.   Very little of the heat can be lost from the chip body in this short time interval by radiation or convection to the air, or by conduction into the tool.

In one investigation, the temperature of the top surface of the chip was measured by a radiation pyrometer.[4]   A low carbon steel was being cut under the following conditions:

cutting speed	160 m min^{-1} (500 ft/min)
feed rate	0.32 mm rev^{-1} (0.013 in/rev)
depth of cut	3 mm (0.125 in)

The temperature of the top surface of the chip was shown to be 335 °C and a very small temperature increase, 2 °C, was recorded as it passed over the contact area on the rake face.

Heat may also be lost from the body of the chip by conduction into the tool through the contact area. It will be shown, however, that, under many conditions of cutting, particularly at high rate of metal removal, the work done in overcoming the feed force, F_f, heats the flow-zone at the under surface of the chip to a temperature higher than that of the body of the chip. Heat then tends to flow into the body of the chip from underneath, and no heat is lost from the body of the chip into the tool by conduction.

Some of the heat generated on the shear plane must be conducted into the approaching work material, which is slightly pre-heated in this way before being sheared to form the chip. If the shear zone were really plane, extending from the cutting edge to the workpiece surface as in the idealised model, the heat conducted back into the work piece would be insignificant unless the shear plane angle were very small. In reality the strain is not confined to a precise shear plane, but takes place in a finite volume of metal, the shape of which varies with the material being cut and the cutting conditions. Quick-stop sections provide evidence that the strained region does not terminate at the cutting edge. *Figure 3.14* shows the flow around the edge of a tool used to cut a very low carbon steel at 16 m min^{-1} (50 ft/min) at a feed rate of 0.25 mm (0.01 in) per rev., while *Figure 5.1a* shows the cutting edge region at higher magnification. The deformed region visibly extends into the work material at the cutting edge to a depth of more than 100 μm (0.004 in). Thus a zone of strained and heated material remains on the new workpiece surface. In many operations, such as turning, much of the metal heated during one revolution of the workpiece is removed on the next revolution, and this portion of the heat is also fed into the chip. However, some of the heat from the deformed layer of work material is conducted back into the workpiece and goes to raise the temperature of the machined part. It is sometimes necessary to remove this heat with a liquid coolant to maintain dimensional accuracy.

The thickness of the deformed layer on the workpiece is very variable. For example, *Figure 5.1b* shows a quick-stop section through the same work material as *Figure 5.1a* but the cutting speed was 110 m min^{-1} (350 ft/min) and the tool was shaped to give reduced contact length. The visibly deformed zone at the work surface below the cutting edge extends to a depth of not more than 20 μm, and heating of the surface must have been correspondingly reduced. Low cutting speeds, low rake angles and other factors which give a small shear plane angle, tend to increase the heat flow into the workpiece. Alloying and treatments which reduce the ductility of the work material, will usually reduce the residual strain in the workpiece.

52

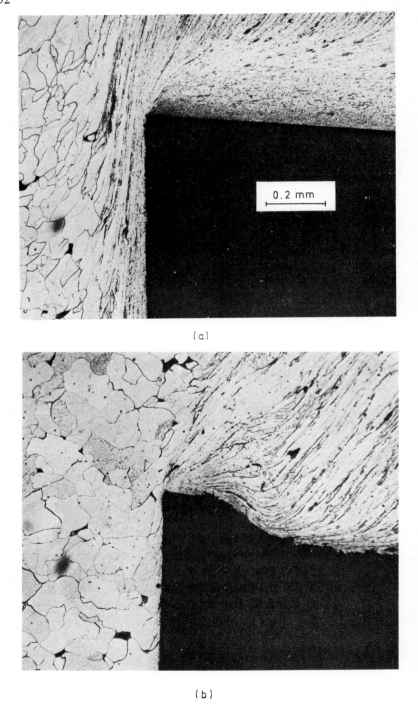

(a)

(b)

Figure 5.1 Metal flow around edge of tool used to cut very low carbon steel (as Figure 3.14). *(a) Using normal tool; (b) using tool with restricted contact on rake face (see* Figure 4.4).

To sum up, most of the heat resulting from the work done on the shear plane to form the chip remains in the chip and is carried away with it, while a small but variable percentage is conducted into the work-piece and raises its temperature. This part of the work done in cutting makes a relatively unimportant contribution to the heating of the cutting tool.

Heat at the tool/work interface

The heat generated at the tool/work interface is of major importance in relation to tool performance, and is particularly significant in limiting the rates of metal removal when cutting iron, steel and other metals and alloys of high melting point. In most publications the generation of heat in this region is treated on the basis of classical friction theory; here the subject is reconsidered in the light of the evidence that *seizure* is a normal condition at the tool/work interface. First it is necessary to discuss in more detail the pattern of strain in the flow-zone, which constitutes the main heat source raising the tool temperature.

Figure 3.10 shows a quick-stop section through the tool/chip interface of a low carbon steel cut by a 6° rake angle tool and *Figure 5.2* shows part of this at a higher magnification. The cutting speed was 153 m min^{-1} (500 ft/min) and the chip thickness ratio was 4:1. The body of the chip was therefore moving over the rake face at 38 m min^{-1} (125 ft/min), while, at the tool face, the two surfaces were not in relative movement. Shear strain (γ) and the units in which it is measured are discussed briefly in Chapter 3 and illustrated diagrammatically in *Figure 3.4* in relation to strain on the shear plane. Values of 2 to 4 for γ on the shear plane are commonly found, and in the present example γ equals approximately 4. The flow-zone was, on the average, 0.075 mm (0.003 in.)

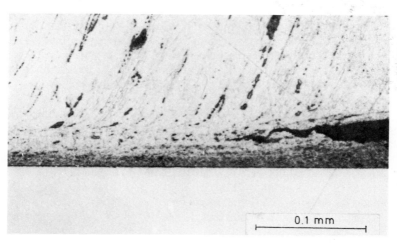

0.1 mm

Figure 5.2 Flow-zone at rake face of tool used to cut very low carbon steel at high speed. Detail of Figure 3.10

thick over most of the seized contact which was 1.5 mm (0.06 in.) long. Shear in the flow-zone can be visualised by considering the change of shape of a unit cube 0.075 mm deep initially situated at the cutting edge on the rake face. When the top of this unit cube had moved to the position where the chip parted from the tool surface, a distance of 1.5 mm, it would have been subjected to a shear strain of 18 or four and one half times as great as the strain on the shear plane. This is not the total extent of strain in the flow-zone, however. Since the bottom of the flow-zone remains anchored to the tool surface, the material in that part very close to the tool surface continues to be subjected to strain indefinitely, as the top of the original unit cube moves off on the underside of the chip when it leaves the tool. Thus the amounts of strain in the flow-zone near the tool surface are very large indeed and cannot be estimated by the methods used for the shear plane.

There is no certain knowledge as to variations in the amount of strain throughout the flow-zone. The usual methods of measuring strain, such as marking a grid on the surface and measuring its change of shape after deformation, are impossible to apply in such a small volume of metal subjected to such extremes of strain. 'Natural markers', structural features in the metal such as grain boundaries and plastic non-metallic inclusions, are drawn out in the flow-zone so nearly parallel with the tool surface that their angle of inclination is too small to be measured as the tool surface is approached (*Figure 5.2*). Photomicrographs of the quick-stop sections provide the best information about the strain pattern in the flow-zone, and from these the thickness of the zone can be measured and a qualitative assessment made of its general character. It is clear, however, that the amounts of strain in the flow-zone are normally several orders of magnitude greater than on the shear plane. The amounts of strain are far outside the range encountered in normal laboratory mechanical testing, where fracture occurs at very much lower strains. The ability of metals and alloys to withstand such enormous shear strains in the flow-zone without fracture, must be attributed to the very high compressive stresses in this region which inhibit the initiation of cracks, and cause the re-welding of such small cracks as may be started or already existed in the work material before machining. For example, the holes in highly porous powder metal components are often completely sealed on the under surface of the chip and on the machined surface where these areas have passed through the flow-zone. It has been shown (*Figures 4.8* and *4.9*) that the compressive stress at the rake face decreases as the chip moves away from the edge, and, when the compressive component of the stress can no longer inhibit the formation of cracks, the chip separates from the tool, its under surface being formed by fracture either at the tool surface or at points of weakness within the flow-zone.

The *rate* of shear strain within the flow-zone is also very high. In the example given (*Figure 3.10*) the mean strain of 18 took place in 2.4 ms, so that the mean rate of strain was 7.5×10^3 s^{-1}. As with the *amount* of strain, what little evidence there is concerning the variation in strain rate throughout the flow-zone comes from photomicrographs of quick-stop sections. These suggest that, although the strain rate may be constant across the thickness of the flow-zone in some areas, there are also areas of 'dead metal' or regions where the strain rate is very low, and others where the rate is much higher than the mean. For example, in *Figure 5.1a* there may be dead metal on the rake face close to the cutting edge, but at the cutting edge itself, where the new surface is being formed, the strain rate appears very high.

A reasonably good estimate of the total amount of heat generated at the tool-work interface can be made from the force and chip thickness measurements. For a zero rake angle tool the heat generated, Q, is

$$Q = \frac{F_f V_c}{J} \qquad (5.2)$$

The temperature of the body of the chip can be calculated with considerable accuracy, equation 5.1, because there is little error involved in assuming even distribution of strain rate across the shear plane, and neglecting heat losses during the short time interval involved. This simplification is not possible in calculation of temperatures in the flow-zone for three reasons:

1. Energy distribution in the flow-zone may be very non-uniform and the data from which to calculate it are unreliable because of the extremes of strain, strain-rate etc.
2. The thickness of the flow-zone, and the amounts of metal passing through, are not accurately known.
3. The heat losses from the flow zone may be large and difficult to calculate.

Many attempts have been made to calculate temperatures and temperature gradients on the rake face of the tool, and progress has been made towards more realistic models and the elimination of sources of error.[5, 6] Even for a two-dimensional section through a tool used for orthogonal cutting, there must be considerable doubt about the accuracy of the values of temperature calculated. Although quantitative estimates of temperature by calculation are uncertain, it is helpful in understanding many aspects of tool life and machinability, to consider the general character of the flow-zone as a heat source.

The material of which the flow-zone is composed is continuously changing. New material is fed in near the cutting edge and is continuously sheared as it flows over the rake face until it leaves the tool as a thin layer on the under surface of the chip. Thus, unlike the material in the body of the chip, it is being continuously sheared, and progressively heated by the work done, and the temperature can therefore be expected to increase as it flows away from the cutting edge.

The temperature increase depends on the amount of work done and on the quantity of metal passing through the flow-zone. The thickness of the flow-zone provides some measure of the latter, and the thinner the flow-zone the higher the temperature would be for the same amount of work done. The thickness varies considerably with the material being

cut from more than 100 μm (0.004 in) to less than 12 μm (0.0005 in).
It tends to be thicker at low speeds, but does not vary greatly with the
feed.   In general the flow-zone is very thin compared with the body of
the chip – commonly of the order of 5% of the chip thickness.   Since
the work done at the tool rake face is frequently about 20% of the
work done on the shear plane, much higher temperatures are found in
the flow-zone than in the chip body, particularly at high cutting speed.
There is very little knowledge at present of the influence of factors such
as work material, tool geometry, or tool material on the flow-zone thickness, and this is an area in which research is required.

The temperature in the flow-zone is influenced strongly, also, by heat
loss by conduction.   The heat is generated in a very thin layer of metal
which has a large area of metallic contact both with the body of the
chip and with the tool.   Since the temperature is higher than that of
the chip body, particularly in the region well back from the cutting edge,
the maximum temperature in the flow-zone is reduced by heat loss into
the chip.   After the chip leaves the tool surface, that part of the flow-
zone which passes off on the under surface of the chip, cools very
rapidly to the temperature of the chip body, since cooling by metallic
conduction is very efficient.   The increase in temperature of the chip
body due to heat from the flow-zone is slight because of the relatively
large volume of the chip body.

The conditions of loss of heat from the flow-zone into the tool are
different from those at the flow-zone/chip body interface because heat
flows continuously into the same small volume of tool material.   It has
been demonstrated (Chapter 3) that the bond at this interface is often
completely metallic in character, and, where this is true, the tool will be
effectively at the same temperature as the flow-zone material at the surface of contact.   The tool acts as a heat sink into which heat flows
from the flow-zone and a stable temperature gradient is built up within
the tool.   The amount of heat lost from the flow-zone into the tool
depends on the thermal conductivity of the tool, the tool shape, and any
cooling method used to lower its temperature.   **The heat flowing into the
tool from the flow-zone raises its temperature and this is the most important factor limiting the rate of metal removal when cutting the higher
melting point metals.**

Heat flow at the tool clearance face

The flow-zone on the rake face is an important heat source because the
chip is relatively flexible and the compressive stress forces it into contact
with the tool over a long path.   In some cases it is possible to reduce
the heat generated by altering the shape of the tool to restrict the length
of contact, *Figure 4.4*.   The same objective is achieved on the clearance
face of the tool by the clearance angle, *Figure 2.2*, which must be large
enough to ensure that the freshly cut surface is separated from the tool
face and does not rub against it.   The heat generated by deformation of

the new surface, *Figure 5.1,* is dissipated by conduction into the work-piece and has little heating effect on the tool.

The work is a more rigid body than the chip, and the feed force cannot deflect it to maintain contact with the tool if a reasonable clearance angle of 5–10° is used. With certain types of tooling, for example form tools or parting-off tools, such a large clearance angle would seriously weaken the tool or make it too expensive, and clearance angles as low as 1° are employed. With such small clearance angles there is a risk of creating a long contact path on the clearance face which becomes a third heat source, similar in character to the flow-zone on the rake face. Even with normal clearance angles, prolonged cutting results in 'flank wear', in which a new surface is generated on the tool more or less parallel to the direction of cutting. The work material is often seized to this 'wear land' as to the rake face of the tool, *Figure 3.5,* and when the worn surface is long enough, this region becomes a serious heat source. Generation of high temperatures in this region is usually followed immediately by collapse of the tool.

Heat in the absence of a flow-zone

Tool temperatures where the heat source is a flow-zone in the work material seized to the tool, have been considered at length because this condition commonly exists where tool life is a problem. Two other conditions are discussed briefly, where the flow-zone is absent from the tool/work interface.

Where a built-up edge is formed there is usually a flow-zone but different in character and further from the tool surface, *Figure 3.15,* and is confined to a restricted region rather than spread out over the rake face. A different type of temperature distribution in the tool can be expected, with lower temperatures, and heat conducted into the tool only in a narrow region close to the tool edge. In most cases, the built-up edge which is formed when cutting steels and other poly-phase materials, disappears as the cutting speed is raised, and is replaced by a flow-zone of the type which has been considered.

Where the conditions at the tool/work interface are those of sliding contact, the mode of heat generation is very different. The shearing of very small, isolated metallic junctions provides numerous very short-lived temperature surges at the interface. This is likely to give rise to a very different temperature pattern at the tool-work interface compared with that of the seized flow-zone. It seems probable that the mean temperatures will be lower, because a much smaller amount of work is required to shear the isolated junctions, but localised high temperatures could also occur at any part of the interface. There is likely to be more even temperature distribution over the contact area than under conditions of seizure. The direct effects of heat generated at sliding contact surfaces would have to be taken into consideration in a complete account of metal cutting. When considering the range of cutting conditions under which problems of industrial machining arise, the effects of heat from areas of sliding contact can probably be neglected without seriously distorting the understanding of the machining process.

Methods of tool temperature measurement

The difficulty of calculating temperatures and temperature gradients near the cutting edge, even for very simple cutting conditions, gives emphasis to the importance of methods for measuring temperature. This has been an important objective of research and some of the experimental methods explored are now discussed.

TOOL-WORK THERMOCOUPLE

The most extensively used method of tool temperature measurement employs the tool and the work material as the two elements of a thermo-couple.[7,8]   The thermo-electric e.m.f. generated between the tool and work piece during cutting is measured using a sensitive millivoltmeter. The hot junction is the contact area at the cutting edge, while an electrical connection to a cold part of the tool forms the cold junction. The tool is electrically insulated from the machine tool (usually a lathe). The electrical connection to the rotating workpiece is more difficult to make, and various methods have been adopted, a form of slip-ring often being used. Care must be taken to avoid secondary e.m.f. sources such as may arise with tipped tools, or short circuits which may occur if the swarf makes a second contact with the tool, for example on a chip breaker. The e.m.f. can be measured and recorded during cutting, and, to convert these readings to temperatures, the tool and work materials, used as a thermocouple, must be calibrated against a standard couple such as chromel-alumel. Each tool and work material used must be calibrated.
   There are several sources of error in the use of this method. The tool and work materials are not ideal elements of a thermo-couple — the e.m.f. tends to be low and the shape of the e.m.f.-temperature curve to be far from a straight line. It is doubtful whether the thermo-e.m.f. from a stationary couple, used in calibration, corresponds exactly to that of the same couple during cutting when the work material is being severely strained. Quite a large number of test results have been reported for different tool and work materials. All show a large increase in temperature as the cutting speed is raised, and temperatures over 1 000 °C have been recorded. The precise significance of the measured temperatures is in doubt because of the very steep temperature gradients known to exist at the interface. Under these conditions, the measured e.m.f. may represent a mean value for the temperature, or possibly the lowest temperature at the interface.
   Errors arising from uncertain calibration of the thermocouple can be partially eliminated by using two different tool materials to cut the same bar of work material, simultaneously, under the same conditions.[9] The

e.m.f. between the two tools is measured. If it can be assumed that the temperatures at the interfaces of the two tools are the same, then the temperature can be determined from the measured e.m.f. by calibration of a thermocouple of the two tool materials. Few results have been reported using this method and, at best, a mean temperature at the contact area can be determined.

INSERTED THERMOCOUPLES

Measurement of the *distribution* of temperature in the tool has been the objective of much experimental work. A simple but very tedious method is to make a hole in the tool and insert a thermocouple in a precisely determined position close to the cutting edge.[10] This must be repeated many times with holes in different positions to map the temperature gradients. The main error in this method arises because the temperature gradients near the edge are very steep, and holes large enough to take the thermocouple overlap a considerable range of temperature. If the hole is positioned very precisely, this may be a satisfactory method for comparing the tool temperature when cutting different alloys, but the method is not likely to be satisfactory for determining temperature distribution.

RADIATION METHODS

Several methods have been elaborated for measurement of radiation from the heated areas of the tool. In some machining operations, the end clearance face is accessible for observation, for example in planing a narrow plate. The end clearance face may be photographed, using film sensitive to infra-red radiation. The heat-image of the tool and chip on the film is scanned using a micro-photometer, and from the intensity of the image the temperature gradient on the end face of the tool is plotted.[11] The results show that the temperature on this face is at a maximum at the rake surface at some distance back from the cutting edge, while the temperature is lower near the cutting edge itself. This method gives information only about the temperature on exposed surfaces of the tool.

A more refined but difficult technique using radiant heat measurement involves making holes, either in the work material or in the tool to act as windows through which a small area (e.g. 0.2 mm dia.) on the rake surface of the tool can be viewed.[4] The image of the hot spot is focussed onto a PbS photo-resistor which can be calibrated to measure the temperature. By using tools with holes at different positions, a map can be constructed showing the temperature distribution on the rake face. A few results using this method have been published showing temperatures when cutting steel at high speeds using carbide tools. These show very high temperatures (up to 1 200 °C) and steep gradients with the highest temperature well back from the cutting edge on the rake face. This method requires very exacting techniques to achieve a single temperature gradient.

STRUCTURAL CHANGES IN HIGH SPEED STEEL TOOLS

Much more information may be obtained using a method described in 1973.[12]   The temperature gradients in three dimensions throughout the part of the tool near the cutting edge, are estimated by observing the structural changes in high speed steel tools under cutting conditions where the temperature is raised over 600 °C.   This temperature is reached when machining the higher melting point metals and alloys at relatively high cutting speeds.   Above 600 °C high speed steels are rapidly 'over-tempered' — the hardness after heating decreases and the structures pass through a series of changes which can be followed by micro-examination after polishing and etching.   *Figure 5.3a* shows a section across the edge of a high speed steel tool after cutting a very low carbon steel at 183 m min⁻¹ (600 ft/min) and a feed rate of 0.25 mm (0.010 in) per rev. for a time of 30 s.   The polished section was etched in Nital (a 2% solution of HNO_3 in alcohol) to reveal the structural changes.

The darkened area forming a crescent below the rake face some distance behind the cutting edge, defines the volume of metal heated above 650 °C during cutting.   The structural changes within this region are shown at higher magnification in *Figure 5.4*.   There are distinct modifications to the structure at approximately 50 °C intervals which permit the estimation of temperature at any position to within ± 25 °C.   The light area at the centre of the crescent on the rake face is a region in which the temperature was above 900 °C.   In this region the transformation temperature of the tool steel was exceeded, the structure became austenitic, and this small volume of steel was re-hardened when cutting stopped because it cooled very rapidly to room temperature.   The changes in the tool steel can be followed by the use of micro-hardness tests.   There is a rapid drop in hardness in those parts which have been heated between 650 °C and 850 °C, and an increase in regions heated to still higher temperatures, above the transformation point.   The micro-hardness method is more tedious than the metallographic assessment of temperature distribution, but it is a useful check, and in some cases, may be more accurate.   To determine temperatures, the particular grade of high speed steel used must be calibrated by a series of tests in which the fully hardened steel is reheated at known temperatures and times, followed by structural investigations and hardness measurements on the reheated pieces.

Figure 5.3b is a temperature contour map based on the structural changes observed in *Figure 5.3a*.   Temperature maps can be constructed also by preparing sections parallel to the tool rake face.   *Figures 5.5a* and *b* show an etched section parallel to the rake face at 12 μm (0.0005 in) below the original surface, and the corresponding temperature map.   The cutting conditions were the same as in *Figure 5.3*; *Figure 5.3a* is thus a section along the line *A-B* in *Figure 5.5a*.   A complete three-dimensional model of temperature distribution can be constructed by a

(a)

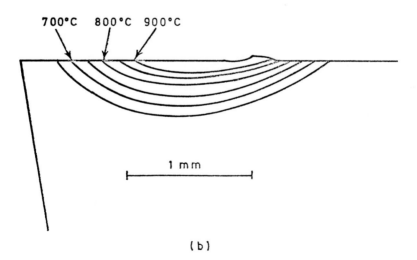

700°C 800°C 900°C

1 mm

(b)

Figure 5.3 (a) Section through tool used to cut very low carbon steel at high speed. Etched in Nital to show heat-affected region. (b) Temperature contours derived from structural changes in (a)[12]

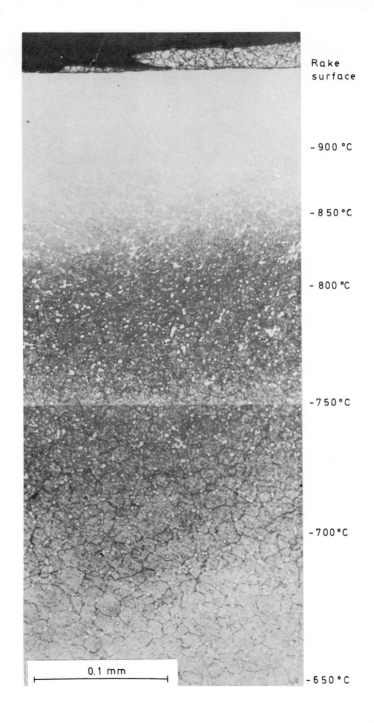

Rake surface

- 900 °C

- 850 °C

- 800 °C

- 750 °C

- 700 °C

0.1 mm

- 650 °C

Figure 5.4   Structural changes and corresponding temperatures in high speed steel tool[1][2]

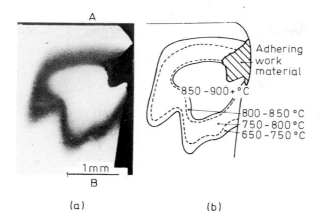

*Figure 5.5 (a) Rake face of tool used to cut very low carbon steel at high speed.
Etched to show heat-affected region. (b) Temperature contours derived from structural
changes in (a)*[13]

series of sections either parallel to or normal to the cutting edge. The
major limitations of this method are that it can be used only on steel
tools, and only under cutting conditions where relatively high temperatures are generated.

Measured temperature distribution in tools

Using the metallographic method, a number of investigations of temperature distribution in cutting tools have been made.[13]   Results of both
theoretical importance and practical interest have been demonstrated,
although insufficient work has been done to enable generalisations to be
drawn valid for all metal cutting.   The results refer to continuous turning
operations, using a 6° positive rake angle tool at a constant depth of cut
(1.25 mm).   Temperature distribution in tools used to cut different work
materials is dealt with in relation to machinability in Chapter 7.   Here,
the effects of a number of parameters are discussed on the basis of a
series of tests using one work material — a very low carbon steel (0.04% C)
of high purity.

Figure 5.6 shows sections through tools used to cut this steel at a feed
rate of 0.25 mm (0.010 in) per rev., at cutting speeds from 91 to 213 m min⁻¹
(300 to 700 ft/min) for a cutting time of 30 s.   The maximum temperature on the rake face of the tool rose as the cutting speed increased, while the hot spot stayed in the same position.   Even at the
maximum speed a cool zone (below 650 °C) extended for 0.2 mm from
the edge, while the maximum temperature, approximately 1 000 °C, is at
a position just over 1 mm from the edge.   This demonstrates the very

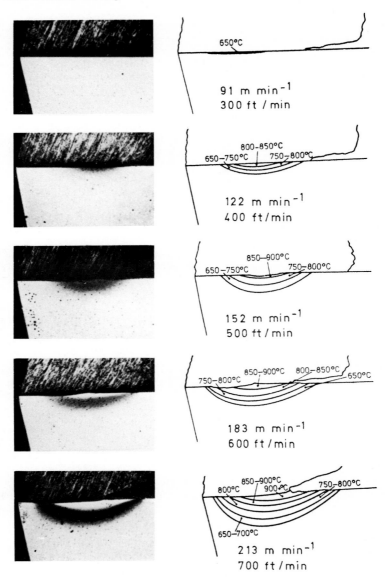

Figure 5.6 Temperature contours in tools used to cut very low carbon steel at a feed rate of 0.25 mm (0.010 in) per rev., and at speeds shown for cutting time of 30 s[13]

high temperatures and very steep temperature gradients which can be pre-sent in the tool at the rake face, when the heat source is a thin flow-zone on the under surface of the chip. As has already been argued, where the flow-zone is metallurgically bonded to the tool surface, there is no barrier to heat conduction, and the temperature at the rake surface of the tool is essentially the same as that in the flow-zone at any position on the interface. The temperature in the flow-zone a few micrometres from the interface may be somewhat higher.

The cool edge of the tool is of great importance in permitting the tool to support the compressive stress acting on the rake face. There is no experimental method by which the stress distribution can be measured under the conditions used in these tests, but the general character of stress distribution, with a maximum at the cutting edge, has already been discussed (*Figures 4.8* and *4.9*). As shown diagrammatically in *Figure 5.7,* the maximum compressive stress is supported by that part of the tool which remains relatively cool under the conditions of these tests.

Heat must flow from the hot spot towards the cutting edge, but this region is cooled by the continual feeding in of new work material. In this series of tests, when the cutting speed was raised above 213 m min^{-1} (700 ft/min), the heated zone approached closer to the edge and the tool failed, the tool material near the edge deforming and collapsing as it was weakened. As the tool deformed, *Figure 5.8a,* the clearance angle near the edge was eliminated, and contact was established between tool and work material down the flank, forming a new heat source, *Figure 5.8b.* Very high temperatures were quickly generated at the flank, the tool was heated from both rake and flank surfaces and tool failure was sudden and catastrophic.

The wear on the clearance face of cutting tools often takes the form of a more or less flat surface – the 'flank wear land' – over which the clearance angle is eliminated (e.g. *Figure 3.5*). This can become a heat

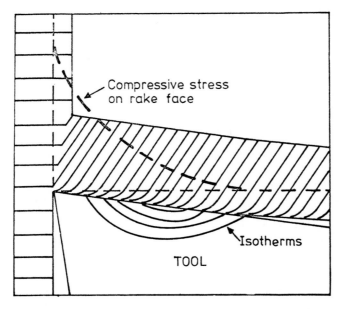

Figure 5.7 Temperature and stress distribution in tool used to cut very low carbon steel[13]

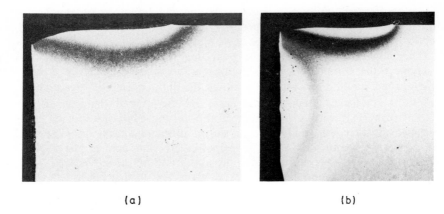

(a)   (b)

Figure 5.8   Stages in tool failure (a) high temperature region spreads to cutting edge   (b) second high temperature region generated at worn clearance face

source causing catastrophic failure if it is allowed to become too large, and it is often recommended that tools be re-ground or replaced when the wear land on the tool reaches some maximum depth, e.g. 0.75 mm or 1.5 mm (0.03 or 0.06 in).

When cutting this low carbon steel, the temperature gradients were established quickly. *Figure 5.6* shows the gradients after 30 s cutting time and they were not greatly different when cutting time was reduced to 10 s. Further experimental work is required to determine temperatures after very short times, such as are encountered in milling.

A few investigations have been made on the influence of feed rate on temperature gradients. *Figure 5.9* shows temperature maps in tools used to cut the same low carbon steel at three different feed rates – 0.125, 0.25, and 0.5 mm (0.005, 0.010, and 0.020 in) per rev. The maximum temperature increased as the feed rate was raised at any cutting speed. The influence of both speed and feed rate on the temperature gradients is summarised in *Figure 5.10,* in which the temperature on the rake face of the tool is plotted against the distance from the cutting edge, for three different speeds at the same feed rate and for three feed rates at the same speed.

The main effect of increasing feed appears to be an increase in length of contact between chip and tool, with extension of the heated area further from the edge and deeper below the rake face, accompanied by an increase in the maximum temperature. The heat flow back toward the cutting edge was increased, and this back flow of heat gradually raised the temperature of the edge, as the feed was increased. There was, however, only a relatively small change in the distance from the edge of the point of maximum temperature when the feed was doubled, and the temperature at the edge was only slightly higher. The highly stressed region near the cutting edge must extend further from the edge as the feed is raised, and this, together with the small increase in edge temperature, sets a limit to the maximum feed which can be employed.

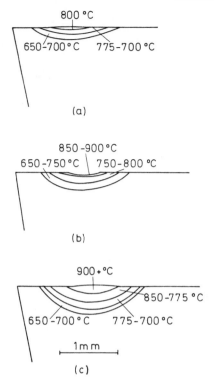

Figure 5.9   Influence of feed rate on temperatures in tools used to cut iron at feed rates (a) 0.175 mm (0.005 in)   (b) 0.25 mm (0.010 in)   (c) 0.5 mm (0.020 in) per rev.

To understand tool behaviour, the temperature distribution in three dimensions must be considered.   This is illustrated by *Figure 5.5* which shows the temperature map of the rake face of a tool used to cut the same steel.   The low temperature region is seen to extend along the whole of the main cutting edge, including the nose radius.   The light area in the centre of the heated region (*Figure 5.5a*) is the part heated above 900 °C, while the sharply defined dark area extending from the end clearance face, is a layer of work material filling a deformed hollow on the tool surface.   The high temperature region is displaced from the centre line of the chip towards the end clearance face.   This is because the tool acts as a heat sink into which heat is conducted from the flow-zone.   Higher temperatures are reached at the end clearance because the heat sink is missing in this region.   With the design of tool used for this test, the only visible surface of the tool heated above red heat was thus at the end clearance face just behind the nose radius.   Tool failure had begun at this position because the compressive stress was high where an unsupported edge was weakened by being heated to nearly 900 °C.

Temperature distribution in tools is thus influenced by the geometry of the tool, for example by the rake angle, by the angle between end and

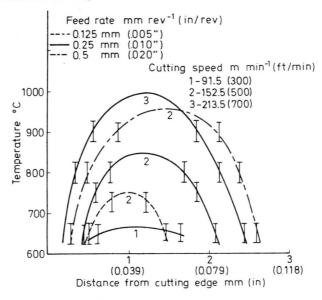

Figure 5.10 Temperature distribution on rake face of tools used to cut very low carbon steel, different speeds and feed rates[13]

side clearance faces or by the nose radius. An understanding of temperature distribution would make possible more rational design of tools and this is an area requiring investigation. While present knowledge if inadequate for *calculation* of temperature gradients in the flow-zone, when temperature distribution can be established experimentally for several sets of conditions, the temperatures in tools of different shape and thermal conductivity could be treated mathematically as a heat flow problem.

The pattern of temperature which has been described for tools used to cut low carbon steel — the cool edge and a high temperature region well back on the rake face — occurs in tools after cutting many steels, including carbon and austenitic stainless steel. The presence of alloying elements has a very large effect on the cutting speed required to produce any temperature. This temperature pattern, however, is *not* a general characteristic of all metal cutting. The characteristics of temperature distribution in tools used to cut different metals are described in Chapter 7 where it is shown that this can have a fundamental influence on the limits to the rate of metal removal for these metals.

Conclusions

A major objective of this chapter is to explain the role of heat in limiting the rate of metal removal when cutting the higher melting point metals. In Chapter 4, experimental evidence is given demonstrating that the *forces* acting on the tool *decrease,* rather than increase, as the cutting speed is raised, and there is no reason to think that the *stresses* on the tool increase with cutting speed. Temperatures at the tool/work interface do, however, increase with cutting speed and it is this rise in temperature which sets the ultimate limit to the practical cutting speed. The most important heat source responsible for raising the temperature of the tool has been identified as the flow-zone where the chip is seized to the rake face of the tool. The amount of heat required to raise the temperature of the very thin flow-zone may represent only a small fraction of the total energy expended in cutting, and the volume of metal heated in the flow-zone may vary considerably. Therefore, there is no direct relationship between cutting forces or power consumption and the temperature near the cutting edge.

Very high temperatures at the tool/work interface have been demonstrated. The existence of temperatures over 1 000 °C at the interface is not obvious to the observer of the machining process, since the high temperature regions are completely concealed, and it is rare to see any part of the tool even glowing red. The thermal assault on the rolls and dies used in the hot working of steel appears much more severe, but the tool materials used for these processes are quite inadequate for metal cutting. It is in order to achieve economic rates of metal removal by machining that the most advanced and sophisticated alloys and materials have been developed as cutting tools.

References

1. TAYLOR, F.W., *Trans. A.S.M.E.*, **28**, 31 (1907)
2. NICOLSON, J.T., *The Engineer*, **99**, 385 (1905)
3. CHILDS, T.H.C. and ROWE, G.W., *Reports on Progress in Physics*, **36**, 3, 225 (1973)
4. LENZ, E., *S.M.E. 1st International Cemented Carbide Conference, Dearborn*, Paper No. MR 71-905 (1971)
5. LOEWEN, E.G. and SHAW, M.C., *Trans. A.S.M.E.*, **76**, 217 (1954)
6. CHAO, B.T. and TRIGGER, K.J., *Trans. A.S.M.E.*, **80**, 311 (1958)
7. HERBERT, E.G., *Proc. Inst. Mech. Eng.*, **1**, 289 (1926)
8. BRAIDEN, P.M., *Proc. Inst. Mech. Eng.*, **182** (3G) 68 (1968)
9. PESANTE, M., *Proceedings of Seminar on Metal Cutting O.E.C.D.*, Paris 1966, 127 (1967)
10. KÜSTERS, K.J., *Industrie Anzeiger*, **89**, 1337 (1956)
11. BOOTHROYD, G., *Proc. Inst. Mech. Eng.*, **177**, 789 (1963)
12. WRIGHT, P.K. and TRENT, E.M., *J.I.S.I.*, **211**, 364 (1973)
13. SMART, E.F. and TRENT, E.M., *Int. J. Prod. Res.*, **13**, 3, 265 (1975)

6

CUTTING TOOL MATERIALS

The development of tool materials for cutting applications has been accomplished very largely by practical men. It has been of an evolutionary character and parallels can be drawn with biological evolution. Millions of people are daily subjecting metal cutting tools to a tremendous range of environments. The pressures of technological change and economic competition have imposed demands of increasing severity. To meet these requirements, new tool materials have been sought and a very large number of different materials has been tried. The novel tool materials (corresponding to genetic mutations) which have been proved by trials, are the products of the persistent effort of thousands of craftsmen, inventors, technologists and scientists — blacksmiths, engineers, metallurgists, chemists. Those tool materials which have survived and are commercially available today, are those which have proved fittest to satisfy the demands put upon them in terms of the life of the tool, the rate of metal removal, the surface finish produced, the ability to give satisfactory performance in a variety of applications, and the cost of tools made from them. The agents of this 'natural selection' are the machinists, foremen, tool room craftsmen, tooling specialists and buyers of the engineering factories, who effectively decide which of all the potential tool materials shall survive.

To reconstruct the whole history of these materials is not possible because so many of the unsuccessful 'mutants' have disappeared without trace or almost so. Patents and the back numbers of engineering journals contain records of some of these. For example *The Engineer* for April 13, 1883, records a discussion at the Institution of Mechanical Engineers where Mr. W.F. Smith described experiments with chilled cast iron tools, which had shown some success in competition with carbon steel. Fuller accounts are available of the work which led to the development of those tool materials which have been an evolutionary success. It is clear that these innovations were made by people who had very little to guide them in the way of basic understanding of the conditions which tools were required to resist at the cutting edge. The simplest concepts of the requirements such as that the tool material should be hard, able to resist heat, and tough enough to withstand impact without fracture, were all that was available to those engaged in this work. It is only in retrospect that a reasoned, logical structure is beginning to emerge to explain the performance of successful tool materials and the failure of other contenders. In this chapter the performance of the main groups of present-day cutting tool materials is discussed. As far as possible the properties of the tool materials are related, on the one hand to their com-position and structure and, on the other, to their ability to resist the tem-peratures and stresses discussed in the previous chapters, and the interact-ions with the work material which wear the tool. The objective is an improved understanding of the performance of existing tools in order to provide a more useful framework of knowledge to guide the continuing evolution of tool materials and the design of tools.

One consequence of this evolutionary form of development has been that the user is confronted with an embarrassingly large number of tool materials from which to select the most efficient for his applications.[1,2] In 1972 in the UK 30 suppliers of high speed steels offered a total of 205 different grades, while 49 suppliers of cemented carbides offered 441 grades. A smell of the magical activities of the legendary Wayland Smith lingers in such trade names as 'Super Hydra' or 'Spear Mermaid'. Many competing commercial alloys are essentially the same, but there is a very large number of varieties which can be distinguished by their composition, properties or performance as cutting tools. These varieties can be grouped into 'species' and groups of 'species' can be related to one another, the groups being the 'genera' of the cutting tool world.

In present-day machine shop practice, the vast majority of tools come from two of these 'genera' — high speed steels and cemented carbides. The other main groups of cutting tool materials are carbon (and lower alloy) steels, cast cobalt-based alloys, ceramics and diamond. The proce-dure adopted here is to consider first high speed steels and factors govern-ing their performance. Cemented carbides are then dealt with similarly, and these two classes of tool material are used to demonstrate general features of the behaviour of tools subjected to the stresses and tempera-tures of cutting. Most of this discussion relates to the behaviour of tools when cutting cast iron and steel, because these work materials have been much more thoroughly explored than others, but, where possible, the per-formance of tools when cutting other materials is described. Finally the use of other classes of tool material is dealt with, but less extensively.

High speed steel

This class of tool steel was developed just before the year 1900, the incentive being increased productivity of the machine shop. The low speed at which steel could be machined had become a severe handicap to its rapidly expanding use in engineering. The tool material available for metal cutting since the beginning of the industrial revolution had been carbon tool steel, and, before the high speed steels, the only significant innovation had been the 'self-hardening' steels initiated by Robert Mushet in England in the 1860s.[3] For use in metal cutting, carbon tool steel was hardened by quenching it in water from a temperature between 750 and 835 °C ('cherry red heat') followed by a tempering treatment between 200 and 350 °C. The 'self-hardening' tool steels were heated in the same temperature range for hardening, but did not need to be quenched in water, air cooling sufficed, and this was an advantage, particularly for large tools. Mushet's development had been achieved by alloying steels with a percentage of tungsten and manganese, while chromium additions were later shown to improve the air-hardening qualities. These tools gave a modest improvement in the speed at which steel could be cut compared with carbon steel tools – for example an increase from 5 to 8 m min^{-1} (16 to 26 ft/min) – but the more expensive self-hardening steels were not very widely adopted by craftsmen with conservative traditions.

The hardening procedures had been ossified by centuries of experience of heat treatment of carbon tool steels. Generations of very competent and highly skilled blacksmiths had established that tools were made brittle by hardening from temperatures above 'cherry-red heat', and this constraint was naturally applied to the heat treatment of self-hardening alloy steels for more than 20 years after their introduction. The story of how revolutionary changes in properties and performance of alloy tool steels was accomplished by modifying the conventional heat treatment, is told by Fred W. Taylor in his Presidential address in 1906 to the A.I.M.E. 'On the Art of Cutting Metals'.[4] This outstanding paper deserves to be read as an historical record of the steps by which the engineer, Taylor, and the metallurgist, Maunsel White, developed the high speed steels, as part of what was probably the most thorough and systematic programme of tests in the history of machining technology. The composition was further developed on the basis of most extensive tool testing, to make full use of the advantages conferred by the new heat treatment. By 1906 the optimum composition was given as:

C	–	0.67%
W	–	18.91%
Cr	–	5.47%
Mn	–	0.11%
V	–	0.29%
Fe	–	Balance

The optimum heat treatment consisted in heating to just below the solidus (the temperature where liquid first appears in the structure – about 1250–1290 °C), cooling in a bath of molten lead to 620 °C, and then to room temperature. This was followed by a tempering treatment just below 600 °C. The tools treated in this way were capable of machining steel at 30 m min^{-1} (99 ft/min) under Taylor's standard test conditions. This was nearly four times as fast as when using the self-hardening steels and six times the cutting speed for carbon steel tools.

At the Paris exhibition in 1900 the Taylor-White tools made a dramatic impact by cutting steel while 'the point of the tool was visibly red hot'. High speed steel tools revolutionised metal cutting practice, vastly increasing the productivity of machine shops and requiring a complete revision of all aspects of machine tool construction. It was estimated that in the first few years, engineering production in the USA had been increased by $ 8 000 m through the use of $ 20 m worth of high speed

steel. Professor Nicolson at the Manchester Municipal School of Technology in 1905 compared the advent of high speed steels with the passing
of the horse and the discarding of reciprocating in favour of rotary prime
movers.

STRUCTURE AND COMPOSITION

To explain this great improvement in performance, it is necessary to consider first the structures of tool steels. The very high hardness of tool
steels is the result of heat treatment which arranges the atoms in a structure known as *martensite*. In martensite the layers of iron atoms are
restrained from slipping over one another by dispersion among them of
the smaller carbon atoms in a formation which forces the iron atoms
out of their normal positions, and locks them into a highly rigid but
unstable structure. This is achieved by quenching, or cooling very rapidly,
from over 730 °C. Although the structure is essentially unstable, the hardness of martensite does not change measurably at room temperature even
after hundreds of years. If a hardened carbon tool steel is reheated
(*tempered*) at a temperature as low as 200 °C, however, the carbon atoms
start to move from their unstable positions, and the steels pass through a
series of transformations, losing hardness and strength, but increasing in
ductility. *Figure 6.1* shows a *tempering curve* for a carbon tool steel –
the hardness at room temperature after reheating for 30 minutes at temperatures up to 600 °C. The tool steel relaxes to a more stable condition
and the high hardness can be restored only by quenching from about
750 °C. If the quenching temperature is raised above 750 °C there is no
further increase in hardness, but the grain size of the steel becomes much
larger, and the steel becomes brittle – the tool edge fractures readily under
impact. In the fully hardened condition carbon tool steels have a maximum hardness of approximately 950 HV. *Figure 6.2a* shows a martensitic
structure as seen in an optical microscope.
   Typical analyses of a carbon tool steel and a high speed steel of today
are:

	C	W	Cr	V	Mn
carbon tool steel	0.9%	–	–	–	0.6%
high speed steel	0.75	18.0	4.0	1.0	0.6

Such a high speed steel when fully heat treated has a hardness of 850
HV – rather lower than that of many carbon steels. *Figure 6.2b* is a
photomicrograph of the structure, the bulk of which, (the *matrix*) consists
of martensite. The alloying elements, tungsten and vanadium, tend to
combine with carbon to form very strongly bonded carbides with the
compositions Fe_3W_3C and V_4C_3 and the former can be seen in the structure of *Figure 6.2b* as small, rounded, white areas a few micrometres (microns)
across. These micrometre-sized carbide particles play an important part in
the heat treatment. As the temperature is raised, the carbide particles

74

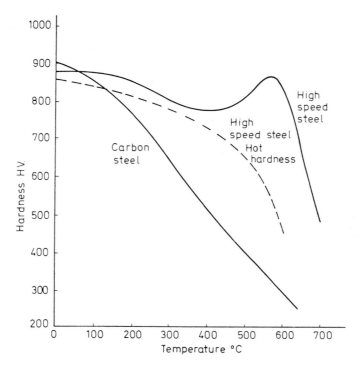

Figure 6.1   Tempering curves for carbon and high speed steel

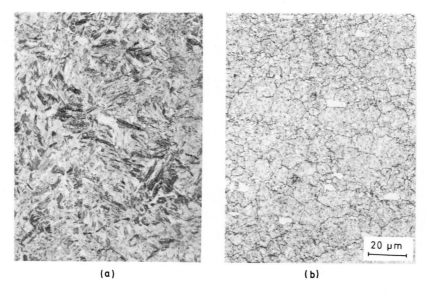

(a)   (b)

Figure 6.2   Micro structures of hardened tool steel   (a) carbon steel; (b) high speed
steel

tend to be dissolved, the tungsten, vanadium and carbon going into solution in the iron, but even up to the melting point some of these particles remain intact, and their presence prevents the grains of steel from growing. It is for this reason that high speed steel can be heated to temperatures as high as 1 290 °C, without becoming coarse grained and brittle.

These carbide particles are harder than the martensitic matrix in which they are held; typical figures are

$$Fe_3W_3C \quad - \quad 1\ 150\ HV$$
$$V_4C_3 \quad - \quad 2\ 000\ HV$$

However, they constitute only about 10 to 15% by volume of the structure, and have a minor influence on the properties and performance of the tools. The vital role in producing the outstanding behaviour of high speed steel is played by carbide particles formed during the tempering operation, and which are only about one thousandth of the size of those visible in *Figure 6.2b*. *Figure 6.1* shows a typical tempering curve for a high speed steel. At first, as with carbon steel, the hardness begins to drop, but over 400 °C it begins to rise again and, after tempering between 500 °C and 600 °C, hardness is often higher than before tempering. With further increase in tempering temperature the hardness falls off rapidly. The *secondary hardening* after tempering at about 560 °C is caused by the formation within the martensite of extremely small particles of carbide. Much of the tungsten and vanadium taken into solution in the iron during the high temperature treatment before hardening, are retained in solution during cooling to room temperature. On re-heating to 400–600 °C they come out of solution and precipitate throughout the structure in the form of extremely numerous carbide particles less than 0.01 μm across. Up to 560 °C the particles remain stable for many hours and harden the steel by blocking slip between the layers of iron atoms. At higher temperatures, however, particularly above 650 °C, the particles coarsen rapidly and lose their capacity for hardening the steel matrix. The hardness can then be restored only by repeating the whole heat treatment cycle.

It is only in recent years that the structural changes responsible for secondary hardening in high speed steel have been understood, and there is still controversy over the precise composition and structure of the fine particles. These are much too small to be seen in an optical microscope, and it was not until sophisticated techniques had been developed with the electron microscope that the vital changes could be observed.[5]

Typical chemical analysis and hardness of the 15 types of high speed steel classified in British Standard Specification 4659 : 1971 are shown in *Table 6.1*.[7] The useful properties of all these grades depend on the development within them of the precipitation-strengthened martensitic structure as a result of high temperature hardening, followed by tempering in the region of 520–570 °C. All of them are softened by prolonged heating to higher temperatures, and no development up to the present has greatly raised the temperature range within which the hardness is retained. A few of these grades, notably BM2 and BT1, are produced in large quantities

Table 6.1 Typical compositions of high speed steels*

Designation	Chemical composition Weight per cent						Hardness HV
	C	Cr	Mo	W	V	Co	Min
BT1	0.75	4	—	18	1	—	823
BT2	0.8	4	—	18	2	—	823
BT4	0.75	4	—	18	1	5	849
BT5	0.8	4	—	19	2	9.5	869
BT6	0.8	4	—	20.5	1.5	12	869
BT15	1.5	4.5	—	12.5	5	5	890
BT20	0.8	4.5	—	22	1.5	—	823
BT21	0.65	4	—	14	0.5	—	798
BT42	1.3	4	3	9	3	9.5	912
BM1	0.8	4	8.5	1.5	1	—	823
BM2	0.85	4	5	6.5	2	—	836
BM4	1.3	4	4.5	6	4	—	849
BM15	1.5	4.5	3	6.5	5	5	869
BM34	0.9	4	8.5	2	2	8	869
BM42	1.05	4	9.5	1.5	1	8	897

*Reproduced from BS4659:1971 by permission of BSI, 2 Park Street, London W1A 2BS from whom, complete copies can be obtained

as general purpose tools, while the others survive because they answer the requirements of particular cutting applications.   A summary of the role of each of the alloying elements gives a guide to the type of applications for which the different grades are suited.

Tungsten and molybdenum.   The essential metallic element in the first high speed steels, upon which the secondary hardening was based, was tungsten, but molybdenum performs the same functions and can be substituted for it.   Equal numbers of atoms of the two elements are required to produce the same properties, and since the atomic weight of molybdenum is approximately half that of tungsten, the percentage of molybdenum in a BM steel is usually about half that of tungsten in the equivalent BT steel. The heat treatment of molybdenum-containing alloys presents rather more difficulties, but the problems have been overcome, and the molybdenum steels are now the most commonly used of the high speed steels.   There is evidence that they are tougher, and their cost is usually considerably lower.

Carbon.   Sufficient carbon must be present to satisfy the bonding of the strongly carbide-forming elements (vanadium, tungsten and molybdenum). An additional percentage of carbon is required, which goes into solution at high temperature and is essential for the martensitic hardening of the matrix.   Precise control of the carbon content is very important.   The highest carbon contents are in those alloys containing large percentages of vanadium.

Chromium.   The alloys all contain 4–5% chromium, the main function of which is to provide *hardenability* so that even those tools of large cross-section may be cooled relatively slowly and still form a hard, martensitic structure throughout.

Vanadium.   All the grades contain some vanadium.   In amounts up to 1%, its main function is to reinforce the secondary hardening, and possibly to help to control grain growth.   A small volume of hard particles of V_4C_3 of microscopic size is formed, and these are the hardest constituents of the alloy.   When the steels contain as high as 5% V, there are many more of these hard particles, occupying as much as 8% by volume of the structure, and these play a significant role in resisting wear, particularly when cutting abrasive materials.

Cobalt.   Cobalt is present in a number of the alloys in amounts between 5 and 12%.   The cobalt raises the temperature at which the hardness starts to fall, and, although the increase in useful temperature is relatively small, the performance of tools under particular conditions may be greatly improved.   The cobalt appears to act by restricting the growth of the precipitated carbide particles, while not itself forming a carbide.

PROPERTIES

Hardness at room temperature is much the most commonly measured property of tool materials.   In the UK the Vickers diamond pyramid indentation hardness test (HV) is used, and can conveniently be carried out on

pieces of many shapes and sizes. It is an effective quality control test
and useful as a first indication of the properties of tool steel. All the
hardness measurements given so far were the results from tests carried out
at room temperature, for example, the tempering curves for BM2 and carbon steel in *Figure 6.1*. *Hot hardness tests* carried out at elevated temperatures, could be of more direct significance for cutting tool performance, but the test procedure is more difficult and the results are less reliable. The dashed line in *Figure 6.1* shows the results of one set of hardness tests on fully heat treated BM2 made at the temperature indicated on
the graph. There is a continuous fall in hardness with increasing temperature, with no peak at 550 °C, and although the hardness is dropping
rapidly, it is still nearly 600 HV at this temperature. *Figure 6.3*[8] shows
the hot hardness of six high speed steels with different room temperature
hardness, at temperatures up to 550 °C. The results show that, for a

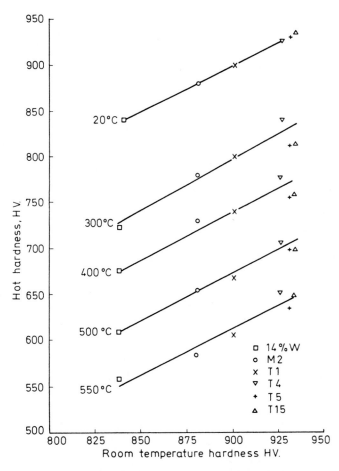

Figure 6.3   Hot hardness vs *cold hardness for 6 high speed steels (F.A. Kirk
et al[8])*

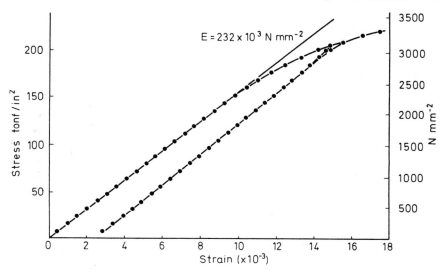

$E = 232 \times 10^3 \text{ N mm}^{-2}$

Figure 6.4   Compression test on type BT1 high speed steel fully heat treated

range of high speed steels, room temperature hardness provides an indication of hot hardness also.   This relationship is true only for materials within the range of the high speed steels.

The compressive strength of fully heat treated steels is of direct relevance to cutting tool performance.   *Figure 6.4* shows the stress-strain curve in compression for a BT1 high speed steel.[9]   The yield stress is very high — 2 270 N mm^{-2} (147 tonf/in^2) — and even at room temperature there is considerable plastic deformation (e.g. 3–5%) before failure. Plastic deformation is accompanied by rapid strain hardening.   *Figure 6.4* shows that when the load was removed after 0.2% plastic strain and the specimen was reloaded, the yield stress had been raised to 3 000 N mm^{-2} (195 tonf/in^2).   There is little information on the effect of temperature on the compressive strength, but *Figure 6.5* shows a comparison between a high speed steel and other tool materials in a relatively high temperature range.[10]   The property measured is the proof stress after 5% reduction in length in compression. A 5% proof stress of 770 N mm^{-2} (50 tonf/in^2) is supported up to 740 °C by the high speed steel, compared with 480 °C for carbon tool steel and 1 000 °C for a cemented carbide. The ability to withstand a high compressive stress at a temperature 250 °C higher is the major factor in the superiority of high speed steel over carbon tool steel for cutting tools.

The breaking strength of high speed steels can be measured using a bend test, and is reported as the maximum tensile stress at the surface of the bar before fracture.   High strengths are recorded, typically in the range 3 500–4 000 N mm^{-2} (230–270 tonf/in^2), and this test was at one time used as a measure of the ability of the tools to resist fracture in service.   At present it is considered that impact tests, using an unnotched bar or a bar with a 'C-notch', are a better measure of *toughness,* and

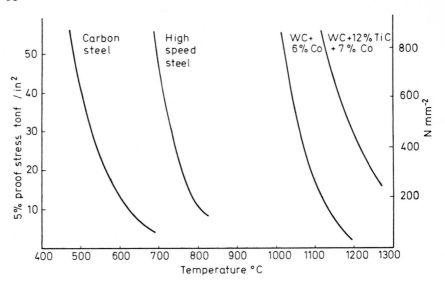

Figure 6.5 Hot compression tests – 5% proof stress of tool steels and cemented carbide[10]

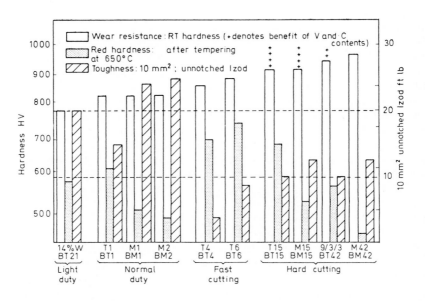

Figure 6.6 Hardness and unnotched impact tests on a selection of high speed steels (F.A. Kirk et al[8]*)*

although inadequate, they seem to be the best tests available. Wide variations in the impact strength of the same grade of high speed steel can result from modifications to the heat treatment, and also from variations in 'quality'. The highest toughness is achieved by ensuring that those carbide particles visible under the microscope are fine and evenly dispersed. For fully heat-treated steels, however, there is an inherent difference in toughness between steels of different composition. *Figure 6.6* compares the toughness as measured by the Izod impact test, of a number of the standard grades, and also shows the hardness at room temperature of the fully heat-treated steels, and their hardness after over-tempering to 650 °C.[8] The toughness of the cobalt-containing steels is markedly lower than that of others.

One of the main areas in which the experience of the specialist is valuable is in selection of the optimum grade and heat treatment to secure maximum tool life and metal removal rate for each particular application. Inadequate toughness will lead to fracture of the tool, giving a short and erratic tool life. The other factors controlling tool life with high speed steel tools are now considered.

TOOL LIFE

For satisfactory performance the shape of the cutting tool edge must be accurately controlled and is much more critical in some applications than in others. Much skill is required to evolve and specify the optimum tool geometry for many operations, to grind the tools to the necessary accuracy and to inspect the tools before use. This is not merely a question of measuring angles and profiles on a macro-scale, but also of inspecting and controlling the shape of the edge on a very fine scale, within a few tenths of a millimetre of the edge, involving such features as burrs, chips or rounding of the edge.

In almost all industrial machining operations the action of cutting gradually changes the shape of the tool edge so that in time the tool ceases to cut efficiently, or fails completely. The criterion for the end of tool life is very varied – the tool may be reground or replaced when it fails and ceases to cut; when the temperature begins to rise and fumes are generated; when the operation becomes excessively noisy or vibration becomes severe; when dimensions or surface finish of the workpiece change or when the tool shape has changed by some specified amount. Often the skill of the operator is required to detect symptoms of the end of tool life, to avoid the damage caused by total tool failure.

The change of shape of the tool edge is very small and can rarely be observed adequately with the naked eye. The skilled tool setter with a watchmaker's lens can see a significant glint from a worn surface, but a binocular microscope with a magnification of at least X 30 is needed even for preliminary diagnosis of the character of tool wear in most cases. The worn surfaces of tools are usually covered by layers of the work material which partially or completely conceal them.   To study the wear of high speed tools it has been necessary to prepare metallographic sections through the worn surfaces, usually either normal to the cutting edge or parallel with the rake face.   The details which reveal the character of the wear process are at the worn surface or the interface between tool and adhering work material, and the essential features are obscured by rounding of the edge of the polished section unless special metallographic methods are adopted.   Sections shown here were mostly prepared by (1) mounting the tool in a cold setting resin in vacuum, (2) carefully grinding the tool to the required section, using much coolant; (3) lapping on metal plates with diamond dust; (4) polishing on a vibratory polishing machine using a nylon cloth with one micron diamond dust.

Observations have shown that the shape of the tool edge may be changed by *plastic deformation* as well as by *wear*.   The distinction is that a wear process always involves some loss of material from the tool surface, though it may also include plastic deformation locally, so that there is no sharp line separating the two.   The use of the term *wear* is, in the minds of many people, synonymous with *abrasion* – the removal of small fragments when a hard body slides over a softer surface, typified by the action of an abrasive grit in a bearing.   There are, however, other processes of wear between metallic surfaces, and a mechanism of *metal-transfer* is frequently described in the literature.   This term is used for an action in which very small amounts of metal, often only a relatively small number of atoms, are transferred from one surface to the other, when the surfaces are in sliding contact, the transfer taking place at very small areas of actual metallic bonding.   Both of these mechanisms may cause wear on cutting tools, but they are essentially processes taking place at sliding surfaces.   There are parts of the tool surface and conditions of cutting, where the work material slides over the tool as in the classical friction model, but, as argued in Chapter 3, it is characteristic of most industrial metal cutting operations that the two surfaces are seized together, and under conditions of seizure there is no sliding at the interface.   Wear under conditions of seizure has not been studied extensively, and investigations of cutting tool wear are in an uncharted area of tribology.   The wear and deformation processes which have been observed to change the shape of high speed steel tools when cutting steel and other high melting point metals are now considered.[11]

Superficial plastic deformation by shear at high temperature

When cutting steel and other high melting-point materials at high rates of speed and feed, a characteristic form of wear is the formation of a *crater*, a hollow in the rake face some distance behind the cutting edge, shown diagrammatically in *Figure 6.7*.   The crater is located at the hottest part of the rake surface, as demonstrated by the method of temperature estimation described in Chapter 5 and illustrated in *Figure 5.3*.   In *Figure 5.3a* the beginnings of the formation of a crater can be observed.   The small ridge just beyond the hottest part on the rake face consists of tool

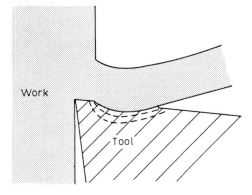

Figure 6.7   Cratering   wear   in   relation   to   temperature   contours

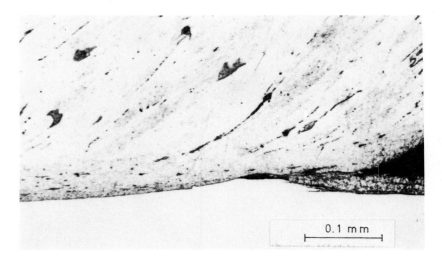

Figure 6.8   Section   through   rake   face   of   tool.   As Figure 5.3, *back   of   crater   in high   speed   steel   tool   after   cutting   iron*[11]

material sheared from the hottest region and piled up behind. This is
shown at higher magnification in *Figure 6.8.*   In this case the work material was a very low carbon steel, the cutting speed was 183 m min⁻¹
(600 ft/min), and the cutting time was 30 s. The maximum temperature
at the hottest position was approximately 950°C.

The steel chip was strongly bonded to the rake face, and this metallographic evidence demonstrates that the stress required to shear the low
carbon steel in the flow-zone, at a strain rate of at least 10^4 s⁻¹, was
high enough to shear the high speed steel where the temperature was
about 950°C. Shear tests on the tool steel used in this investigation gave
a shear strength of 110 N mm⁻² (9 tonf/in²) at 950°C at a strain rate of
0.16 s⁻¹. It is unexpected to find that nearly pure iron can exert a stress

high enough to shear high speed steel.   This is possible because (1) the yield stress of high speed steel is greatly lowered at high temperature, and (2) the rate of strain of the low carbon steel in the flow-zone is very high, while the rate of strain in the high speed steel, although it cannot be estimated, can be lower by several orders of magnitude.   In both materials the yield stress increases with the strain rate.

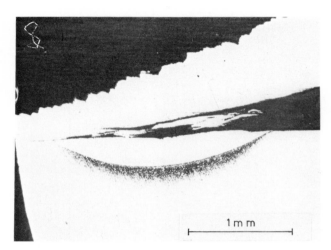

1 m m

Figure 6.9   Crater in high speed steel tool used to cut austenitic stainless steel. Etched to show heat affected region below crater filled with work material. (P.K. Wright[11])

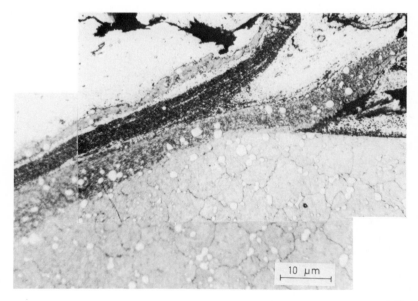

10 μm

Figure 6.10   End of crater after cutting austenitic stainless steel.   (P.K. Wright[11])

This wear mechanism has been observed on tools used to cut many steels. *Figures 6.9* and *6.10* show cratering of high speed steel tools used to cut austenitic stainless steel. The shearing away of successive layers of tool steel is particularly well seen in *Figure 6.10* at the rear end of the crater. It occurs also when cutting titanium and its alloys, and nickel and its alloys. When cutting the stronger alloys it occurs at much lower cutting speeds than when cutting the commercially pure metals. This is because the shear stress is higher and the tool is therefore sheared at lower interface temperatures. However, it always occurs at the regions of highest temperature at the tool-work interface, and when cutting nickel alloys, it has been observed at the cutting edge (*Figure 6.11*) where high temperatures are generated with these materials. It also occurs on a tool flank when the tool is severely worn with accompanying high temperatures.

This is a rapid-acting wear mechanism, forming deep craters which weaken the cutting edge so that the tool may be fractured. This wear mechanism may not be frequently observed under industrial cutting conditions, but it is a form of wear which sets a limit on the speed and feed rates which can be used when cutting the higher melting point metals with high speed steel tools. It is unlikely that this wear mechanism will be observed when cutting copper-based or aluminium-based alloys, since, with these lower melting point materials, both the temperatures generated and the shear yield stresses are very much lower.

Plastic deformation under compressive stress

The compressive stress acting on the rake face is discussed in Chapter 4. Under laboratory test conditions the stress is a maximum at, or close to, the cutting edge (*Figure 4.8*), and it is a reasonable assumption that this is true for most conditions of cutting, at least in the absence of a built-up edge. When the stress is very high, the tool edge may be deformed downward as shown in *Figure 6.12*, a section through a tool used to cut cast iron. This is a *deformation* rather than a *wear* process, since no tool material is removed, but it results in increased tool forces, and brings into play or accelerates wear processes which reduce the life of the tool. As the cutting speed is raised, the stress near the edge probably does not increase and may even fall, but the temperature rises, the yield stress of the tool material is reduced, and deformation starts when the tool is critically weakened. Deformation of the tool edge, together with the previously discussed shearing away of the tool surface, are two mechanisms which often set a limit to the speed and feed rate which can be used. When cutting steel the cutting edge is relatively cool and may be undeformed by the high compressive stress, while a crater is being formed by a relatively low shear stress at the hottest position on the rake face (*Figure 6.9*). When cutting nickel-based alloys, deformation of the tool edge is more serious because of high temperatures at the edge.

Tools are more likely to be damaged by deformation when the hardness of the work material is high and it is this mechanism which limits

the maximum workpiece hardness which can be machined with high speed
steel tools, even at very low speed where temperature rise is not important.   An upper limit of 350 HV is often considered as the highest hardness at which practical machining operations using high speed steel tools,
can be carried out on steels, though steel as hard as 450 HV may be cut
at very low speed.   Deformation often leads to sudden failure of the tool
by fracture or by localised heating.

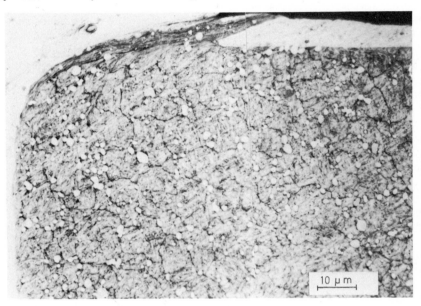

*Figure 6.11   Shearing of high speed steel tool at cutting edge after cutting nickel-
based alloy.   (P.K. Wright[11])*

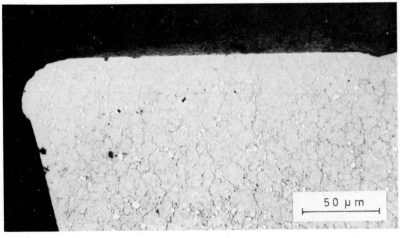

*Figure 6.12   Deformation of cutting edge of tool used to cut cast iron.   (P.K.
Wright[11])*

Diffusion wear

The metallographic evidence shows that conditions exist during cutting where diffusion across the tool/work interface is probable. There is metal to metal contact and temperatures are high enough for appreciable diffusion to take place. Thus tools may be worn by metal and carbon atoms from the tool diffusing into and being carried away by, the stream of work material flowing over its surface. There is evidence that appreciable diffusion does occur in regions at high temperature, for example, intermediate layers of a new phase between the tool and work materials are observed under some cutting conditions, and these appear to have been formed by diffusion. The white band in *Figure 6.13*, at the rake face of a tool used to cut cast iron is such a layer.

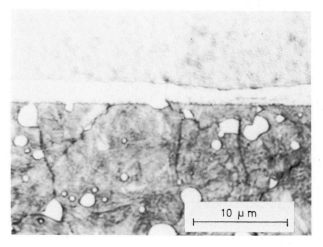

Figure 6.13   White-etching layer formed at rake face of high speed steel tool after cutting steel at medium cutting speed.   (P.K. Wright[11])

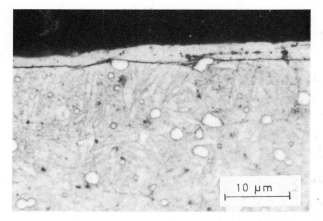

Figure 6.14   Section through rake face of high speed steel tool after cutting steel. Interface characteristic of diffusion wear.   (P.K. Wright[11])

Direct evidence of wear by atomic diffusion cannot be obtained by optical metallography, but the appearance of some worn surfaces is most easily explained by the occurrence of this process. Rates of diffusion increase rapidly with temperature, the rate typically doubling for an increase of the order of 20°C. Therefore evidence for diffusion wear would be sought at the high temperature regions of the seized interface. The rapid form of cratering caused by superficial plastic deformation has been described, but, at somewhat lower cutting speeds, craters form more slowly and there is no evidence of plastic deformation of the tool. *Figure 6.14* is a section through such a worn surface with a thin layer of the work material (steel) seized to the tool. The wear is of a very smooth type and no plastic deformation is observed, the grain boundaries of the tool steel being undeformed up to the surface. The carbide particles were not worn at all, or worn much more slowly, undermined and eventually carried away. Diffusion wear is a sort of chemical attack on the tool surface, like etching, and is dependent on the solubility of the different phases of the tool material in the metal flowing over the surface, rather than on the hardness of these phases. The carbide particles are more resistant because of their lower solubility in the steel work material, whereas there are no solubility barriers to the diffusion of iron atoms from the tool steel into a steel work material. The diffusion wear process is one which has not been seriously considered as a cause of wear, except in relation to metal cutting. The rate of diffusion wear is very dependent on the metallurgical relationship between tool and work material and this is important when cutting different metals such as titanium or copper. It is of more significance for cemented carbide tools than for high speed steel and is considered again in relation to them.

With high speed steel tools used in the usual cutting speed range, rates of wear by diffusion are relatively slow because the interface temperatures are relatively low. At higher speeds and higher temperatures, diffusion is accelerated, but diffusion wear is masked by the plastic deformation which is a much more rapid wear mechanism. Diffusion accounts for the formation of craters at speeds below those at which plastic deformation begins, and is probably the most important wear process responsible for flank wear in the higher speed range. Wear by diffusion depends both on high temperatures, and also on a rapid flow rate in the work material very close to the seized surface, to carry away the tool metal atoms. On the rake face, very close to the cutting edge, there is usually a dead metal region, or one where the work material close to the tool surface flows slowly (*Figure 5.1a*), and this, together with the lower temperature, accounts for the lack of wear at this position. The rate of flow past the flank surface is, however, very high when cutting at high speeds, and wear by diffusion can take place on the flank, although the temperature is very much the same as on the adjacent unworn rake face.

Attrition wear

At relatively low cutting speed, temperatures are low, and wear based on plastic shear or diffusion does not occur. The flow of metal past the

Figure 6.15 Section through cutting edge of high speed steel tool after cutting steel at relatively low speed showing attrition wear. (P.K. Wright[11])

Figure 6.16 Detail of Figure 6.15. (P.K. Wright[11])

cutting edge is more irregular, less stream-lined or laminar, a built-up edge may be formed and contact with the tool may be less continuous. Under these conditions larger fragments, of microscopic size, may be torn inter- mittently from the tool surface, and this mechanism is called *attrition*. *Figures 6.15* and *6.16* show a section through the cutting edge of a high speed steel tool after cutting a medium carbon steel for 30 minutes at 30 m min^{-1} (100 ft/min). A built-up edge remained when the tool was disengaged. The tool edge had been 'nibbled' away over a considerable period of time. Fragments of grains had been pulled away, with some tendency to fracture along the grain boundaries, *Figure 6.16*, leaving a very uneven worn surface. The tool and work materials are strongly bonded together over the whole of the torn surface. Although relatively large fragments were removed, this must have happened infrequently once a stable configuration had been reached, because the tool had been cutting for a long time.

In continuous cutting operations using high speed steel tools, attrition is usually a slow form of wear, but more rapid destruction of the tool edge occurs in operations involving interruptions of cut, or where vibra- tion is severe due to lack of rigidity in the machine tool or very uneven work surfaces. Attrition is not accelerated by high temperatures, and tends to disappear at high cutting speed as the flow becomes laminar. This is a form of wear which can be detected and studied only in metal- lographic sections. Adhering metal often completely conceals the worn surface and, under these conditions, visual measurements of wear on the untreated tool may be misleading.

Abrasive wear

Abrasive wear of high speed steel tools requires the presence in the work material of particles harder than the martensitic matrix of the tool. Hard carbides, oxides and nitrides are present in many steels, in cast iron and in nickel-based alloys, but there is little direct experimental evidence to indicate whether abrasion by these particles does play an important role in the wear of tools. Some evidence of abrasion of rake and flank sur- faces by Ti (C,N) particles are seen in sections through tools used to cut austenitic stainless steel stabilised with titanium. *Figure 6.17* shows a Ti (C,N) particle which had ploughed a groove in the rake face of the tool as it moved from left to right, eventually remaining partially embedded in the tool surface. In this experiment a corresponding steel without the hard particles showed only slightly less wear when cut under the same conditions. It seems doubtful whether, under conditions of seizure, small, isolated hard particles in the work material can make an impor- tant contribution to wear. *Figure 6.17* suggests that they may be quickly stopped, and partially embedded in the tool surface where they would act like microscopic carbide particles in the tool structure.

Abrasion is intuitively considered as a major cause of wear and the literature on the subject often describes tool wear in general as abrasive, but this is an area that requires further investigation for normal conditions

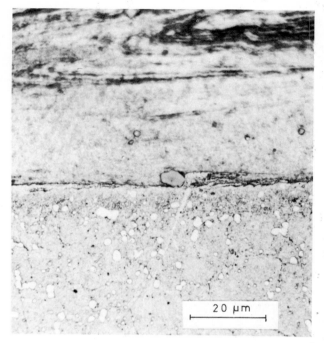

Figure 6.17 Section through rake face of high speed steel tool after cutting austenitic stainless steel containing Ti. Abrasive action by Ti(C,N) particle. (P.K. Wright[11])

of cutting. Where the work material contains greater concentrations of hard particles, such as pockets of sand on the surface of castings, rapid wear by abrasion undoubtedly occurs. In such concentrations the action is like that of a grinding wheel, and the surfaces of castings are treated to remove abrasive material in order to improve tool life.

Wear under sliding conditions

At those parts of the interface where sliding occurs, either continuously or intermittently, other wear mechanisms can come into play and, under suitable conditions, can cause accelerated wear in these regions. The parts of the surface particularly affected are those shown as areas of intermittent contact in *Figure 3.13*. The most frequently affected are those marked *E, H* and *F* in *Figure 3.13*, on the rake and clearance faces. Greatly accelerated wear on the rake face at the position *EH*, and down the flank from *E* where the original work surface crossed the cutting edge, is shown in *Figure 6.18*. The wear mechanisms operating in these sliding regions are probably those which occur under more normal engineering conditions at sliding surfaces, involving both abrasion and metal

(a)   (b)

Figure 6.18   (a) Rake surface and (b) flank of high speed steel tool used to cut steel showing built-up edge and sliding wear at position E.   (P.K. Wright[11])

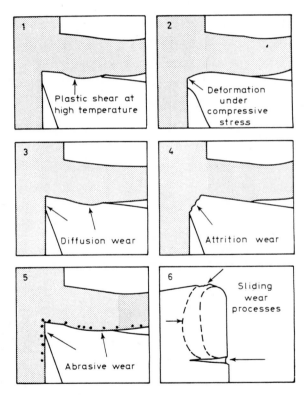

Figure 6.19   Wear mechanism on high speed steel tools

transfer, and greatly influenced by chemical interactions with the surrounding atmosphere.   This is further discussed in relation to the action of cutting lubricants in Chapter 8.

Figure 6.19 summarises the discussion of the wear and deformation processes which have been shown to change the shape of the tool, and to affect tool life when cutting steel, cast iron, and other high temperature metals and alloys, with high speed steel tools. The relative importance of these processes depends on many factors — the work material, the machining operation, cutting conditions, tool geometry, and use of lubricants. In general the first three processes are important at high rates of metal removal where temperatures are high, and their action is accelerated as cutting speed increases. It is these processes which set the upper limit to the rate of metal removal. At lower speeds, tool life is more often terminated by one of the last three — abrasion, attrition or a sliding wear process — or by fracture. Under unfavourable conditions the action of any one of these processes can lead to rapid destruction of the tool edge, and it is important to understand the wear or deformation process involved in order to take correct remedial action.

TOOL-LIFE TESTING

Most data from tool-life testing have been compiled by carrying out simple lathe-turning tests in continuous cutting, using tools with a standard geometry, and measuring the width of the flank wear land and sometimes the dimensions of any crater formed on the rake face. Steel and cast iron have been the work materials in the majority of reported tests. The results of one such test programme, cutting steel with high speed tools over a wide range of speed and three feed rates is given by König[12] and one set of results is summarised in *Figure 6.20*. For crater wear the results are simple, the wear rate being very low up to a critical speed, above which cratering increased rapidly. This critical speed is lowered as the feed is increased. For flank wear also the wear rate increases rapidly at about the same speed and feed as for cratering. It is in this region that the temperature-dependent wear mechanisms control tool life. Below this critical speed range, flank wear rate does not continue low, but often increases to a high value as attrition and other wear mechanisms not dependent on temperature become dominant.

Very high standards of systematic tool testing were set by F.W. Taylor in the work which culminated in the development of high speed steel. The variables of cutting speed, feed rate, depth of cut, tool geometry and lubricants, as well as tool material and heat treatment were studied and the results presented as mathematical relationships for tool life as a function of all these parameters. These tests were all carried out by lathe turning of very large steel billets using single point tools. Such elaborate tests have been too expensive in time and manpower to be repeated frequently, and it has become customary to use standardised conditions, with cutting speed and feed rate as the only variables. The results are presented using what is called *Taylor's equation,* which is

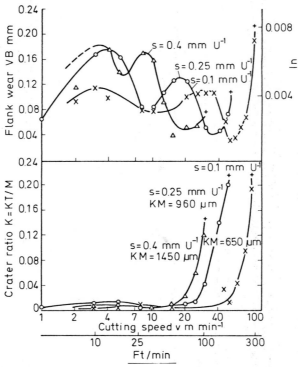

Figure 6.20   *Influence of cutting speed and feed rate on flank and crater wear of high speed steel tools after cutting steel.   Work material $Ck_{55}N$ (AISI C 1055); tool material S 12-1-4-5; depth of cut a = 2 mm (0.08 in); tool geometry a = 8°, γ = 10°, λ = 4°, x = 90°, ϵ = 60°, r = 1 mm; cutting time t = 30 min (Opitz and König[12])*

Taylor's original relationship reduced to its simplest form:

$$VT^n = C \tag{6.1}$$

or
$$\log V = \log C - n \log T \tag{6.2}$$

where V = cutting speed
        T = cutting time to produce a standard amount of flank wear (e.g. 0.75 mm, 0.030 in)
and C and n are constants for the material or conditions used.

   Figure 6.21   shows the results of one set of tests in which the time to produce a standard amount of flank wear is plotted on a logarithmic scale against the cutting speed.[13]   In spite of considerable scatter in individual test measurements, the results often fall reasonably well on a straight line.   From these curves the cutting speed can be read off for a tool life of, for example, 60 minutes (V 60) or 30 minutes (V 30) and work materials are sometimes assessed by these numbers.   Pronounced and significant differences are demonstrated in *Figure 6.21* between the different

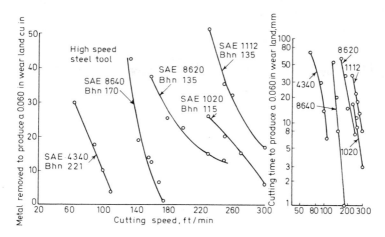

Figure 6.21 'Taylor' tool-life curves. Metal removed and cutting time to produce a 0.060-in (1.5 mm) wear land on a high speed steel tool (18-4-1) vs *cutting time for five annealed steels. Tool angles in degrees: back rake, 0; side rake, 15; side cutting edge angle, 0; end cutting edge angle, 5; relief 5. Nose radius, 0.005 in (0.13 mm); feed, 0.009 in/rev (0.23 mm rev⁻¹); depth of cut 0.062 in (1.57 mm)*[13] *(By permission, from Metals Handbook, 1st Supplement, Copyright American Society for Metals, 1954)*

steels machines and such graphs are often presented as an evaluation of 'machinability'. They should be considered rather as showing one aspect of machinability or of tool performance – i.e. the tool life for a given tool and work material and tool geometry *when cutting in the high speed range.*

The 'Taylor curves' are valid for those conditions in the high speed range where tool life is controlled by the temperature-dependent processes involving deformation and diffusion. It could be implied from *Figure 6.21* that, if cutting speed were reduced to still lower values, the tool life would become effectively infinite – the tools would never wear out. Such extrapolations to lower cutting speeds are not valid. As shown in *Figure 6.20* tool wear rates may increase again at lower speeds because other wear mechanisms come into play. In many practical operations it is not possible to use high cutting speeds – for example along the cutting edge of a drill, when turning or forming small diameters, or when planing or shaping. For many operations the 'Taylor curves' are therefore not suitable for predicting tool life.

CONDITIONS OF USE

An extremely wide range of sizes and shapes of high speed steel tools is in use today. Most of these are in the form of solid tools, consisting entirely of high speed steel, which are shaped by machining close to the final size when the steel is in the annealed condition. After hardening, the final dimensions are achieved by grinding and this operation requires

much skill and experience. There is a considerable difference in the difficulty of grinding between one grade and another, the high vanadium grades
causing most problems, with very high rates of grinding wheel wear.

In some cases the high speed steel is in the form of an insert which is
clamped, brazed or welded to the main body of the tool, which is made
of a cheaper carbon or low alloy steel. This is a tendency which is
increasing, and recent developments include drills where the high speed
steel cutting end is friction welded to a carbon steel shank, and band
saws in which a narrow band of high speed steel, with teeth, is electron-
beam welded to a low alloy steel band. So far, however, the small
'throw-away' clamped tip, commonly used with carbide tools, has not
been widely adopted for high speed steel. Most steel tools are designed
to be reground when worn, so that each tool edge may be reprepared
for use many times.

SUMMING-UP ON HIGH SPEED STEEL TOOLS

High speed steel tools make possible the cutting of steel and other high
melting-point materials at much higher rates of metal removal than can
be achieved with carbon steel tools. This improved performance is made
possible by their retention of hardness and compressive strength to higher
temperatures. These properties are the result of precipitation hardening
within the martensitic structure of these tools steels after a high temperature heat treatment. The correctly heat-treated steels have adequate
toughness to resist fracture in machine shop conditions, and it is the combination of high-temperature strength and toughness which has made this
class of tool material one of the successful survivors in the evolution of
cutting tool materials. The loss of strength and permanent changes in
structure of high speed steel when heated above 600 °C, limit the rate of
metal removal when cutting higher melting-point metals and alloys.

Cemented carbides

Many substances harder than quenched tool steel have been known from
ancient times. Diamond, Corundum and quartzite, among many others
were natural materials used to grind metals. These could be used in the
form of loose abrasive or as grinding wheels, but were unsuitable as metal-
cutting tools because of inadequate toughness. The introduction of the
electric furnace in the last century, led to the production of new hard
substances at the very high temperatures made available. The American
chemist Acheson produced silicon carbide in 1891 in an electric arc
between carbon electrodes. This is used loose as an abrasive and, when
bonded with porcelain, is very important as a grinding wheel material,
but is not tough enough for cutting tools.

Many scientists, engineers and inventors in this period explored the use
of the electric furnace with the aim of producing synthetic diamonds. In
this they were not successful, but Henri Moissan at the Sorbonne made

many new carbides, borides and silicides – all very hard materials with high melting-points. Among these was tungsten carbide which was found to be exceptionally hard and had many metallic characteristics. It did not attract much attention at the time, but there were some attempts to prepare it in a form suitable for use as a cutting tool or drawing die. There are, in fact, two carbides of tungsten – WC which decomposes at $2\,600°C$, and W_2C which melts at $2\,750°C$. Both are very hard and there is a eutectic alloy at an intermediate composition and a lower melting point ($2\,525°C$). This can be melted and cast with difficulty, and ground to shape using diamond grinding wheels, but the castings are coarse in structure, with many flaws. They fracture easily and proved to be unsatisfactory for cutting tools and dies.

In the early 1900s the work of Coolidge led to the manufacture of lamp filaments from tungsten, starting with tungsten powder with a grain size of a few micrometres (microns). The use of the powder route of manufacture eventually solved the problem of how to make use of the hardness and wear resistant qualities of tungsten carbide. In the early 1920s Schröter, working in the laboratories of Osram in Germany, heated tungsten powder with carbon to produce the carbide WC in powder form, with a grain size of a few micrometres (microns).[14,15] This was thoroughly mixed with a small percentage of a metal of the iron group – iron, nickel or cobalt – also in the form of a fine powder. The mixed powders were pressed into compacts which were sintered by heating in hydrogen to above $1\,300°C$. The product was completely consolidated by the sintering process and consisted of fine grains of WC bonded or 'cemented' together by a tougher metal. Cobalt metal was soon found to be the most efficient metal for bonding. These *cemented carbides* have a unique combination of properties which has led to their development into the second major genus of cutting tool materials in use today.

STRUCTURE AND PROPERTIES

Tungsten-carbide is one of a group of compounds – the carbides, nitrides, borides and silicides of transition elements of Groups IV, V and VI of the Periodic Table.[16,17] Of these, the carbides are important as tool materials, and the dominant role has been played by the mono-carbide of tungsten, WC. *Table 6.2* gives the melting point and room temperature hardness of some of the carbides. All the values are very high compared with those of steel. The carbides of tungsten and molybdenum have hexagonal structures, while the others of major importance are cubic. These rigid and strongly bonded compounds undergo no major structural changes up to their melting points, and their properties are therefore stable and unaltered by heat treatment, unlike steels which can be softened by annealing and hardened by rapid cooling.

These carbides are strongly metallic in character, having good electrical and thermal conductivities and a metallic appearance. Although they have only slight ability to deform plastically without fracture at room temperature, the electron microscope has shown that they are deformed by the

Table 6.2

	Melting point °C	Diamond indentation hardness HV
TiC	3 200	3 200
V$_4$C$_3$	2 800	2 500
NbC	3 500	2 400
TaC	3 900	1 800
WC	2 750 decomposes	2 100

NOTE. There are large differences in reported values for all of the carbides. The values given here are representative.

same mechanism as are metals – by movement of dislocations. They are sometimes included in the category of ceramics, and the cemented carbides have been referred to as 'cermets', implying a combination of ceramic and metal, but this term seems quite inappropriate, since the carbides are much closer in character to metals than to ceramics.

Figure 6.22 shows the hardness of four of the more[18,19] important carbides measured at temperatures from 15 °C to over 1 000 °C. All are much harder than steel, and for comparison, the room temperature hardness of diamond on the same scale is 6 000–8 000 HV. The hardness of the carbides drops rapidly with increasing temperature, but they remain much harder than steel under almost all conditions. The very high hardness, and the stability of the properties when subjected to a wide range of thermal treatments, are favourable to the use of carbides in cutting tools.

In cemented carbide alloys, the carbide particles constitute about 55–92% by volume of the structure, and those alloys used for metal cutting normally contain at least 80% carbide by volume. *Figure 6.23* shows the structure of one alloy, the angular grey particles being WC and the white areas cobalt metal. In high speed steels the hard carbide particles of microscopic size constitute only 10–15% by volume of the heat-treated steel (*Figure 6.2b*), and play a minor role in the performance of these alloys as cutting tools, but in the cemented carbides they are the decisive constituents. The powder metal production process makes it possible to control accurately both the composition of the alloy and the size of the carbide grains.

Tungsten carbide–cobalt alloys

This group, technologically the most important, is considered first. These are available commercially with cobalt contents between 4% and 30% by weight – those with cobalt contents between 4% and 12% being commonly used for metal cutting – and with carbide grains varying in size between 0.75 μm and 10 μm across. The performance of carbide cutting tools is very dependent upon the composition and grain size and also upon

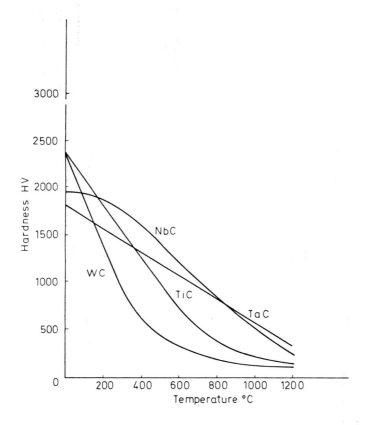

Figure 6.22   Hot hardness tests on monocarbides of four transition elements (after Atkins and Tabor;[18] Miyoshi and Hara[19])

Figure 6.23   Structure of a cemented carbide (WC-Co) of coarse grain size

Table 6.3 Properties of WC–Co alloys

Co%	Mean WC grain size μm	Hardness HV 30	Transverse rupture strength N mm⁻²	Transverse rupture strength tonf/in²	Compressive strength N mm⁻²	Compressive strength tonf/in²	Young's modulus N mm⁻² × 10³	Young's modulus tonf/in² × 10³	Specific gravity
3	0.7	2 020	1 000	65					
	1.4	1 820							
6	0.7	1 800	1 750	113	4 550	295			14.95
	1.4	1 575	2 300	148	4 250	275	630	40.7	
9	0.7	1 670	2 300	148	4 000	260			14.75
	1.4	1 420	2 400	156	4 000	260	588	38.0	
	4.0	1 210	2 770	179					
15	0.7	1 400	2 770	179	3 500	225			14.00
	1.4	1 160	2 600	168			538	34.8	

the general quality of the product. The structure of tungsten carbide–cobalt alloys should show two phases only – the carbide WC and cobalt metal (*Figure 6.23*). The carbon content must be controlled within very narrow limits. The presence of either free carbon (too high a carbon content) or 'eta phase' (a carbide with the composition Co_3W_3C, the presence of which denotes too low a carbon content) results in reduction of strength and performance as a cutting tool. The structures should be very sound showing very few holes or non-metallic inclusions.

Table 6.3 gives properties of a range of WC–Co alloys in relation to their composition and grain size, and *Figure 6.24* shows graphically the influence of cobalt content on some of the properties. Both hardness and compressive strength are highest with alloys of low cobalt content and decrease continuously as the cobalt content is raised.[20] For any composition the hardness is higher the finer the grain size and over the whole range of compositions used for cutting, the cemented carbides are much harder than the hardest steel. As with high speed steels, the tensile strength is seldom measured, and the breaking strength in a bend test or 'transverse rupture strength' is often used as a measure of the ability to resist fracture in service. *Figure 6.24* shows that the transverse rupture strength varies inversely with the hardness, and is highest for the high cobalt alloys with coarse grain size.

In tensile or bend tests, fracture occurs with no measurable plastic deformation, and cemented carbides are, therefore, often characterised as 'brittle' materials in the same category as glass and ceramics. This is not justified because, under conditions where the stress is largely compressive, cemented carbides are capable of considerable plastic deformation before failure. *Figure 6.25* shows stress–strain curves for two cemented carbide alloys in compression, in comparison with high speed steel. The cemented carbide has much higher Young's modulus (E), and higher yield stress than the steel. Above the yield stress, the curve departs gradually from linearity, and plastic deformation occurs accompanied by strain hardening, raising still further the yield stress. The amount of plastic deformation before fracture increases with the cobalt content.[21,22] The ability to yield plastically before failure can also be demonstrated by making indentations in a polished surface. Indentations in a carbide surface, made with a carbide ball, are shown in *Figure 6.26,* the contours of the impressions being demonstrated by optical interferometry. Indentations to a measurable depth can be made before a crack appears, and the higher the cobalt content the deeper the indentation before cracking. It is this combination of properties – the high hardness and strength, together with the ability to deform plastically before failure under compressive stress – which makes cemented carbides based on WC so well adapted for use as tool materials in the engineering industry.

There is no generally accepted test for toughness in tool materials and the terms *toughness* and *brittleness* have to be used in a qualitative way. This is unfortunate because it is on the basis of toughness that the selection of the optimum tool material must often be made. Frequently the best tool material for a particular application is the hardest which has adequate toughness to resist fracture. In practice the solution to this

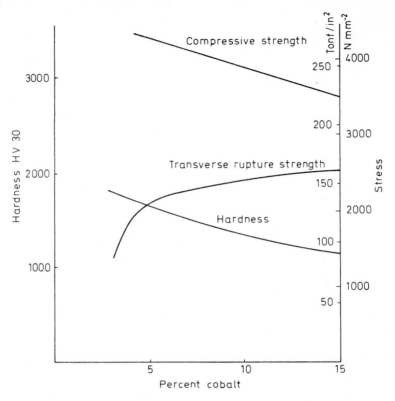

Figure 6.24   Mechanical properties of WC-Co alloys of medium-fine grain size

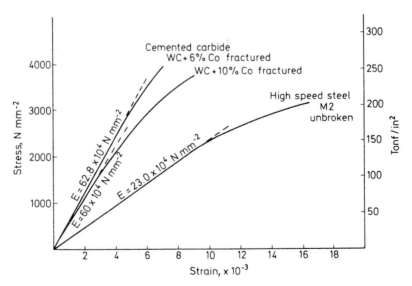

Figure 6.25   Compression tests on two WC-Co alloys compared with high speed steel[10]

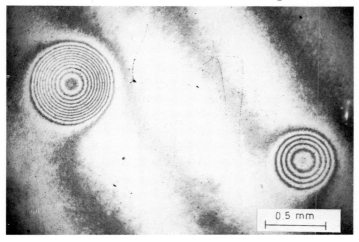

Figure 6.26 Ball indentations in surface of cemented carbide, revealed by optical interferometry

Figure 6.27 Structure of medium-fine grained alloy of WC with 6% Co ('Wimet N')

problem has been achieved by the 'natural selection' process of industrial experience. Just as, for high speed steels, a general purpose grade, BM2, has become established for general machining work, so with the WC–Co alloys, a grade containing 6% Co and a grain size near 2 μm, *Figure 6.27*, has become acknowledged as the grade for the majority of applications for which this range of alloys is suitable, and this grade is made in much larger quantities than the others. If this grade proves too prone to fracture in a particular application, a grade with a higher cobalt content is tried, while, if increased wear resistance is needed, one with a finer grain size or lower cobalt content is recommended.

Table 6.4 Classification of carbides according to use*

Symbol	Broad categories of material to be machined	Designation	Material to be machined	Use and working conditions
P		P 01	Steel, steel castings	Finish turning and boring; high cutting speeds, small chip section, accuracy of dimensions and fine finish, vibration-free operation.
		P 10	Steel, steel castings	Turning, copying, threading and milling, high cutting speeds, small or medium chip sections.
		P 20	Steel, steel castings Malleable cast iron with long chips	Turning, copying, milling, medium cutting speeds and chip sections, planing with small chip sections
	Ferrous metals with long chips	P 30	Steel, steel castings Malleable cast iron with long chips	Turning, milling, planing, medium or low cutting speeds, medium or large chip sections, and machining in unfavourable conditions†
		P 40	Steel Steel castings with sand inclusion and cavities	Turning, planing, slotting, low cutting speeds, large chip sections with the possibility of large cutting angles for machining in unfavourable conditions† and work on automatic machines
		P 50	Steel Steel castings of medium or low tensile strength, with sand inclusion and cavities	For operations demanding very tough carbide: turning, planing, slotting, low cutting speeds, large chip sections, with the possibility of large cutting angles for machining in unfavourable conditions† and work on automatic machines

Direction of increase in characteristic of cut of carbide

- Toughness →
- ← Wear resistance
- Increasing feed →
- ← Increasing speed

M — Ferrous metals with long or short chips and non-ferrous metals

Grade	Material	Application
M 10	Steel, steel castings, manganese steel; Grey cast iron, alloy cast iron	Turning, medium or high cutting speeds. Small or medium chip sections
M 20	Steel, steel castings, austenitic or manganese steel, grey cast iron	Turning, milling. Medium cutting speeds and chip sections
M 30	Steel, steel castings, austenitic steel, grey cast iron, high temperature resistant alloys	Turning, milling, planing. Medium cutting speeds, medium or large chip sections
M 40	Mild free cutting steel, low tensile steel; Non-ferrous metals and light alloys	Turning, parting off, particularly on automatic machines

← Increasing speed | Increasing feed → | ← Wear resistance | toughness →

K — Ferrous metals with short chips, non-ferrous metals and non-metallic materials

Grade	Material	Application
K 01	Very hard grey cast iron, chilled castings of over 85 Shore, high silicon aluminium alloys, hardened steel, highly abrasive plastics, hard cardboard, ceramics	Turning, finish turning, boring, milling scraping
K 10	Grey cast iron over 220 Brinell, malleable cast iron with short chips, hardened steel, silicon aluminium alloys, copper alloys, plastics, glass, hard rubber, hard cardboard, porcelain, stone	Turning, milling, drilling, boring broaching, scraping
K 20	Grey cast iron up to 220 Brinell, non-ferrous metals: copper, brass, aluminium	Turning, milling, planing, boring broaching, demanding very tough carbide
K 30	Low hardness grey cast iron, low tensile steel, compressed wood	Turning, milling, planing, slotting, for machining in unfavourable conditions† and with the possibility of large cutting angles
K 40	Soft wood or hard wood; Non-ferrous metals	Turning, milling, planing, slotting, for machining in unfavourable conditions† and with the possibility of large cutting angles

← Increasing speed | Increasing feed → | ← Wear resistance | toughness →

* Reproduced from ISO513:1975(E) by permission of ISO, Geneva. Copies can be obtained from BSI, 2 Park Street, London W1A 2BS
† Raw material or components in shapes which are awkward to machine: casting or forging skins, variable hardness etc., variable depth of cut, interrupted cut, work subject to vibrations

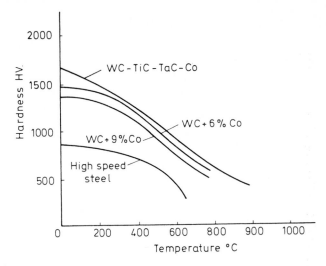

Figure 6.28 Hot hardness of cemented carbides compared with high speed steel

Trial by industrial experience is a very prolonged and expensive way of establishing differences in toughness. Also the absence of a generally recognised, objective toughness test makes it difficult for the user to assess the claims of competing suppliers regarding the qualities of their carbide grades. There is at present no British Standard Specification for cemented carbides. The user must rely on classifications such as that of the International Organisation for Standardisation (ISO)[23], (*Table 6.4*) in which the performance as cutting tools is related to the *relative* toughness and hardness of the different WC–Co alloys, arranged in a series from K01 to K40. It is the responsibility of the individual manufacturer to decide in which category each of his grades shall be placed.

Both hardness and compressive strength of cemented carbides decrease as the temperature is raised (*Figures 6.28* and *6.5*). The comparison of compressive strength at elevated temperatures with that of high speed steel[10] (*Figure 6.5*) shows that a WC–Co alloy with 6% Co withstands a stress of 750 Nmm^{-2} (50 tonf/in^2) at 1 000°C, while the corresponding temperature for high speed steel is 750°C. With cemented carbides the temperature at which this stress can be supported drops if the cobalt content is raised.

The coefficient of thermal expansion is low – about half that of most steels. Thermal conductivity is relatively high: a 6% Co–94% WC alloy has a thermal conductivity, of 80 W/m°C compared with 31 W/m°C for high speed steel. Oxidation resistance at elevated temperatures is poor, oxidation in air becomes rapid over 600°C and at 900°C is very rapid indeed. Fortunately this rarely becomes a serious problem with cutting tools because the surfaces at high temperature are usually protected from oxidation (see Chapter 8).

PERFORMANCE OF TUNGSTEN CARBIDE–COBALT TOOLS

Cemented tungsten carbide tools were introduced into machine shops in the early 1930s and inaugurated a revolution in productive capacity of machine tools as great as that which followed the introduction of high speed steels. Over a period of years the necessary adaptations have been made to tool geometry, machine tool construction, methods of operation and attitude of mind of the machinists, which enable the tools to be used efficiently. When premature fracture was avoided, the carbide tools could be used to cut many metals and alloys at much higher speed and with much longer tool life. There are limits to the rate of metal removal at which there is a reasonable tool life. As with high speed steel tools, the shape of the edge is gradually changed with continued use until the tool no longer cuts efficiently. The mechanisms and processes which change the shape of the edge of tungsten carbide–cobalt tools when cutting cast iron, steel and other high melting-point alloys are now considered, looking first at those which set the limits to the rate of metal removal.[24,25,26]

As with high speed steel tools, the work material is seized to the tool over much of the worn rake and flank surfaces when cutting cast iron and steel at medium and high cutting speeds. The heat source is, therefore, a flow-zone of the same character as that described for steel tools. The first wear mechanism described for high speed steels (shearing away of the tool surface to form a crater) is not normally observed when using carbide tools, their shear strength at high temperature being too great for this to occur.

Plastic deformation under compressive stress

The limit to the rate of metal removal as cutting speed or feed are raised, is often caused by deformation of the tool under compressive stress on the rake face. Carbide tools can withstand only limited deformation, even at elevated temperature, and cracks form which lead to sudden fracture. *Figure 6.29* shows such a crack in the rake face of a tool, this surface being stressed in tension as the edge is depressed. Failure due to deformation is more probable at high feed rates, and when cutting materials of high hardness. Carbide grades with low cobalt content can be used to higher speed and feed rate because of increased resistance to deformation. The improved room temperature hardness of carbides with finer grain size, however, is not an indication of improved resistance of the tools to deformation at elevated temperature.

Deformation can be detected at an early stage in a laboratory tool test by lapping the clearance face optically flat before the test, and observing this face after the test using optical interferometry. Any deformation in the form of a bulge on the clearance face can be observed and measured as a contour map formed by the interference pattern. *Figure 6.30* shows the interference fringe pattern on the clearance face of a carbide tool used to cut steel at high speed and feed. The maximum deformation is

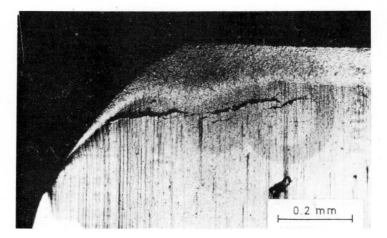

0.2 mm

Figure 6.29   Cracks across the nose of cemented carbide tool deformed during cutting[35]

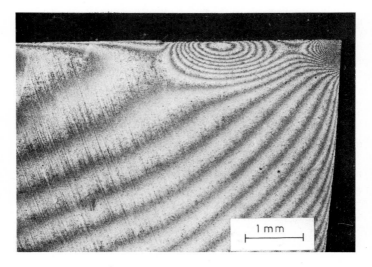

1 mm

Figure 6.30   Deformation on clearance face of cemented carbide tool, after cutting at high speed and feed, revealed by optical interferometry[26]

seen to be at the nose, and this is a common feature when cutting steel and iron. A tool with a sharp nose, or a very small nose radius, starts to deform at the nose and fails, for this reason, at much lower speed than when a large nose radius is used. In many cutting operations the precise form of the nose radius may be critical for the performance of the tool.

The ability of tungsten carbide-cobalt tools to resist deformation at high temperatures is the most important property permitting higher rates of metal removal compared with steel tools.

Diffusion wear

When cutting steel at high speed and feed rate, a crater is formed on the rake face of tungsten carbide–cobalt tools, with an unworn flat at the tool edge (*Figure 6.31*). Measurement of temperature in carbide tools is not possible by the technique used for high speed steels, but the temperature gradient in the carbide tools must be of the same general type because the heat source is of the same character − a thin flow-zone in metallic contact with the rake face of the tool. The crater is in the same position as on high speed steel tools, the deeply worn crater being associated with the high temperature regions, and the unworn flat with the low temperatures near the edge. Sections through the crater in WC–Co tools show no evidence of plastic deformation by shear in the tool material. The carbide grains are smoothly worn through (*Figure 6.32*) with little or no evidence that any particles large enough to be seen under the light microscope are broken away from the tool surface.

There is strong evidence that the wear process in cratering of WC–Co tools is one in which the metal and carbon atoms of the tool diffuse into the work material seized to the surface and are carried away in the chip. An extreme example of this sort of process was observed when studying sections through WC–Co bullet cores embedded in steel plates. These showed a layer in which WC and Co were dissolved in the steel which had been melted at the interface.[27] *Figure 6.33* shows such a section with steel at the top, the cemented carbide at the bottom, and the white, fused layer in between containing partially dissolved, rounded WC grains. The melting point of a eutectic between WC and cobalt or WC and iron is about 1 300 °C and this temperature had been reached in the thin layer

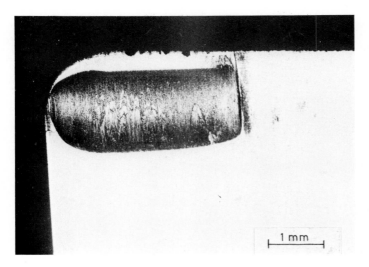

Figure 6.31 Crater on rake face of WC-Co alloy tool after cutting steel at high speed and feed rate

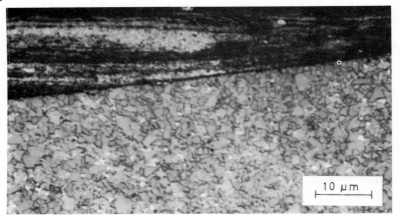

Figure 6.32   Section through crater surface in WC-Co tool and adhering steel, showing interface characteristic of diffusion wear

Figure 6.33   Section through interface of WC-Co bullet core embedded in steel plate

at the interface.   In this case, the speed on impact exceeded 1 000 m s⁻¹
and a very large amount of energy had been converted into heat in a very
short interval of time.   In a paper published in 1952[28] the author
suggested that the cratering of WC-Co tools when cutting steel was also
the result of fusion at the interface at a temperature of 1 300 °C.   With
cutting tools, however, the speeds involved are more than 100 times lower,
and evidence of temperature distribution in high speed steel tools, such
as that in *Figure 5.3* suggests that sustained temperatures at the interface
are considerably lower than 1 300 °C for conditions where cratering first
occurs, probably in the region of 900–1 100 °C.   While these temperatures
are too low for fusion, they are high enough to allow considerable diffusion to take place in the solid state, and the characteristics of the observed
surfaces are consistent with a wear process based on solid phase diffusion.

   Worn carbide tools can be treated in HCl or other mineral acids to
remove adhering steel or iron so that the worn areas of the tool surface
can be studied in detail.   While the acids dissolve the cobalt binder to
some extent they do not attack the carbides, and wear on carbide tools
can thus be examined by a method not possible for high speed steel
tools.   *Figures 6.34* and *6.35* show the worn crater surfaces of WC-Co
alloy tools after cutting steel.   The carbide grains are mostly smoothly
worn through, but sometimes have an etched appearance, as in the large

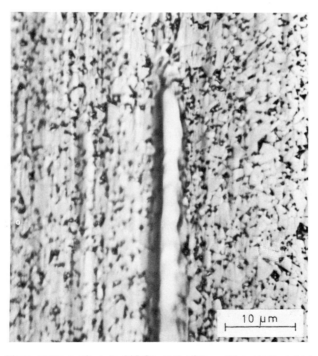

*Figure 6.34   Worn crater surface in WC-Co tool after cutting steel at high speed.
Very smooth surface with ridge characteristic of diffusion wear*

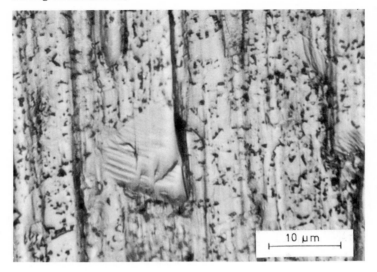

Figure 6.35 Worn crater surface on WC-Co tool, showing large WC grain 'etched' by action of the hot steel at the interface[25]

grain in *Figure 6.35* in which steps etched in the surface are parallel to one of the main crystallographic directions. Smooth ridges are often seen on the surface, originating from large WC grains and running in the direction of chip flow, as in *Figure 6.34*. The smooth polished surfaces are characteristic of areas where the work material flows rapidly immediately adjacent to the surface, while the etched appearance is seen where the temperature is high enough but the flow is less rapid or there is a stagnant layer at the interface.

Cratering wear by diffusion occurs only at relatively high rates of metal removal. This and other features of wear on cemented carbide tools can be conveniently summarised by 'machining charts' such as *Figure 6.36*.[35] The coordinates on the chart are the cutting speed and feed rate plotted on a logarithmic scale, and the diagonal dashed lines are lines of equal rate of metal removal. On each chart are plotted the occurrence of the main wear features for one combination of tool and work material using a standard tool geometry. Two lines in *Figure 6.36* show the conditions under which cratering wear was first detected, and the conditions under which it became so severe as to cause tool failure within a few minutes of cutting. *Figure 6.36* is for a WC–6% Co tool when cutting a medium carbon steel. The occurrence of crater wear by diffusion is a function of both cutting speed and feed rate and an approximately straight line relationship is found to exist over a wide range of cutting conditions. For example the limit to cutting speed caused by rapid cratering was 90 m min⁻¹ (300 ft/min) at a feed rate of 0.25 mm rev⁻¹ (0.01 in/rev), while the critical speed is reduced to 35 m min⁻¹ (120 ft/min) at a feed rate of 0.75 mm rev⁻¹ (0.030 in/rev). These charts and the figures for maximum cutting speeds illustrate the importance of

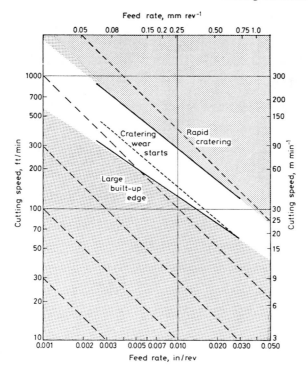

Figure 6.36 *'Machining chart' for WC + 6% Co tool cutting 0.4% C steel with hardness 200 HV*[35]

the cratering-diffusion wear in limiting the rates of metal removal when using WC–Co alloys to cut steel. The maximum rates of metal removal are not much higher than those when using high speed steel tools, and for this reason WC–Co alloys are not often used for cutting steel. They are recommended mainly for cutting cast iron and non-ferrous metals, although they can be used to advantage to give longer tool life when cutting steel at relatively low speeds.

Since diffusion wear is of such importance with carbide cutting tools, it is worth considering this wear mechanism in more detail.[6,25] The rate of diffusion across the interface depends on temperature, but also on many other factors such as the relative sizes of the atoms involved, their chemical interactions, and solubilities of the materials in each other. There must be some solubility for diffusion to take place at all. It has been shown, for example, that 7% of WC can be dissolved in iron at 1 250 °C. Thus the rate of diffusion wear depends on what is sometimes called the 'compatibility' of the materials; large differences in diffusion wear rate occur with different tool and work materials. The rate of wear is more dependent on the chemical properties than on the mechanical strength or hardness of the tool, provided the tool is strong enough to withstand the imposed stresses. It is for this reason that the higher hardness of carbides with fine grain size is not reflected in improved resistance to diffusion wear. In fact coarse-grained alloys are rather more resistant than fine-grained ones of the same composition, but the difference is small.

The rate of diffusion wear depends on the rate at which atoms from the tool diffuse into the work material and consideration is now given to the question – 'which atoms from the tool material are most important?' In the case of high speed steels, the iron atoms from the matrix diffuse into the work until the isolated carbide particles, which remain practically intact, are undermined and carried away bodily. With cemented carbide tools also, the most rapid diffusion is by the cobalt atoms of the carbide bond, and the iron atoms of the work material. The carbide grains, however, are not undermined and carried away for two reasons. First, because the carbide particles are not isolated, but constitute most of the volume of the cemented carbide, supporting each other in a rigid framework. Secondly, because, as cobalt atoms diffuse out of the tool, so iron atoms diffuse in, and iron is almost as efficient in 'cementing' the carbide as is cobalt.

Carbon atoms are small and can move rapidly through iron, but those in the tool are strongly bonded to the tungsten and are not free to move away by themselves. It is the rate of diffusion of tungsten and carbon atoms together into the work material which controls the rate of diffusion wear. This depends not only on the temperature, but also on the rate at which they are swept away – i.e. on the rate of flow of the work material very close to the tool surface – at distances of 0.001–1 μm. Just as the rate of evaporation of water is very slow in stagnant air, so the rate of diffusion wear from the tool is low where the work material is stationary at, and close to, the tool surface. At the flank of the tool, the rate of flow of the work material close to the tool surface is very high (*Figure 5.1a*), and diffusion may be responsible for a high rate of flank wear even when the adjacent rake face surface is practically unworn. In *Figure 3.5* the carbide grains on the flank can be seen to be smoothly worn through. Under conditions of seizure, the smooth wearing through of carbide grains can be regarded as a good indication that a diffusion wear process is involved.

When cutting at relatively high speeds where the flank wear is based on diffusion, the wear rate increases rapidly as the cutting speed is increased. *Figure 6.37* shows a typical family of curves of flank wear against cutting time for cutting steel with carbide tools in the higher range of cutting speeds. In the WC–Co alloys, the percentage of cobalt influences the rate of wear by diffusion, the flank wear rate rising with increasing cobalt content, but, within the range of grades commonly used for cutting, the differences in wear rate are not very great provided the tools do not become deformed.

Attrition wear

As with high-speed steel tools, when cutting at relatively low speeds, where temperature is not high enough for wear based on diffusion or deformation to be significant, attrition takes over as the dominant wear process. The

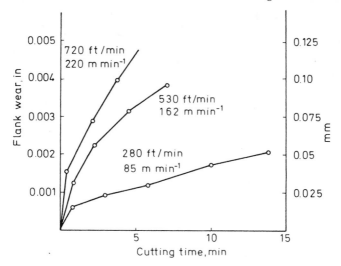

Figure 6.37   Flank wear vs *time for increasing cutting speeds when cutting steel with WC-Co tools where wear is mainly by diffusion*[25]

condition for this is a less laminar and more intermittent flow of the work material past the cutting edge, of which the most obvious indication is the formation of a built-up edge. *Figure 6.36* is typical of the charts for many steels, in that it shows the presence of a built-up edge at relatively low rates of metal removal, dependent on both cutting speed and feed rate. It is below the built-up edge line on the chart that wear is largely controlled by attrition.

During cutting, the built-up edge is continually changing, work material being built on to it and fragments sheared away (*Figure 3.15*). If only the outer layers are sheared, while the part of the built-up edge adjacent to the tool remains adherent and unchanged, the tool continues to cut for long periods of time without wear. For example, under some conditions when cutting cast iron, the built-up edge persists on WC–Co tools to relatively high cutting speeds and feed rates, as shown on the chart, *Figure 6.38*. With grey cast iron the built-up edge is infrequently broken away and tool life may be very long. It is for this reason that WC–Co alloy tools are commonly used for cutting cast iron, and the recommended speeds are those where a built-up edge is formed. The wear rate is low but is of the attrition type, whole grains or fragments of carbide grains being broken away leaving the sort of worn surface shown in the sections *Figures 3.8* and *3.9*.

When cutting steel, however, under conditions where a built-up edge is formed, the edge of a WC-Co alloy tool may be rapidly destroyed by attrition. As with high-speed steel tools, fragments of the tool material of microscopic size are torn from the tool edge, but whereas this is a slow wear mechanism with steel tools, it may cause rapid wear on carbide tools. If the built-up edge is firmly bonded to the tool and is

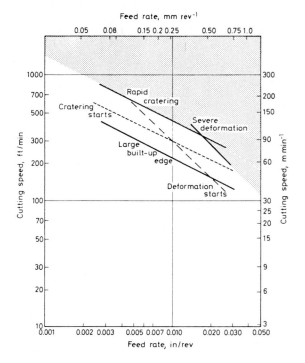

Feed rate, mm rev⁻¹

Figure 6.38 'Machining chart' for WC + 6% Co tool cutting pearlitic flake graphite cast iron[35]

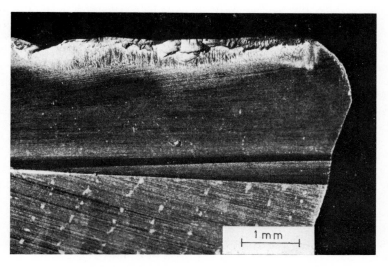

Figure 6.39 Edge chipping of carbide tool after cutting steel at low speed, with a built-up edge

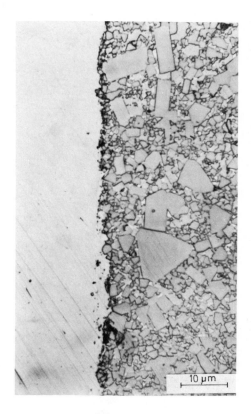

Figure 6.40   Section through flank wear and adhering metal on WC-Co tool used for cutting steel at low speed, showing attrition wear[25]

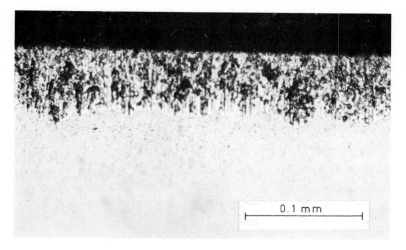

Figure 6.41   Worn flank of cemented carbide tool, adhering steel removed in acid. Surface shows evidence of both diffusion and attrition wear

118

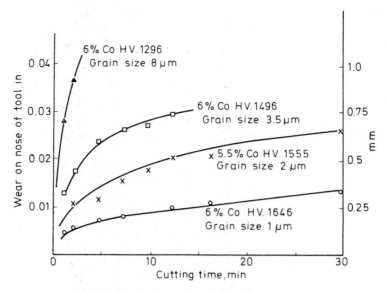

Figure 6.42   Flank wear on WC-Co tools showing influence of carbide grain size when cutting cast iron under conditions of attrition wear

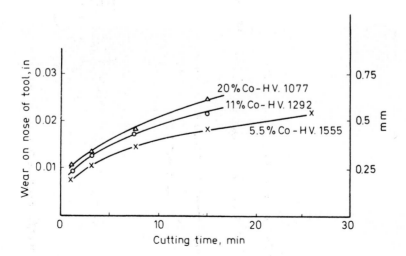

Figure 6.43   Flank wear on WC-Co tools showing influence of cobalt content when cutting cast iron under conditions of attrition wear

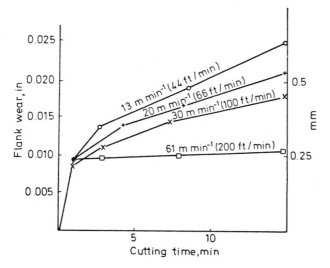

Figure 6.44 Flank wear vs *time for increasing cutting speeds when cutting with WC-Co tools where wear is by attrition — compare* Figure 6.37[25]

broken away as a whole, as frequently happens where cutting is interrupted, relatively large fragments of the tool edge may be torn away as shown in *Figure 6.39*. Where the machine tool lacks rigidity, or the work piece is slender and chatter and vibration occur, the metal flow past the tool may be very uneven and smaller fragments of the tool are removed. *Figure 6.40* shows WC grains being broken up and carried away in the stream of steel flowing over the worn flank of a WC-Co tool used to cut steel at low speed and feed rate.

Fragments are broken away because localised tensile stresses are imposed by the unevenly flowing metal. Steel tools are stronger in tension with greater ductility and toughness and for this reason have greater resistance to attrition wear. This is one of the main reasons why high-speed steel tools are employed, and will continue to be used, for many machining operations at low cutting speeds. Worn surfaces produced by attrition are very rough compared with the almost polished surfaces resulting from diffusion wear. There is, however, no sharp dividing line between the two forms of wear, both operating simultaneously, so that worn surfaces, when adhering metal has been dissolved, often show some grains smoothly worn and others torn away, *Figure 6.41*.

The rate of wear by attrition is not directly related to the hardness of the tool. With WC-Co tools, the most important factor is the grain size, fine-grained alloys being much more resistant than coarse-grained ones. *Figure 6.42* shows the rates of wear of a series of tools all containing 6% cobalt when cutting cast iron in laboratory tests under conditions of attrition wear. The hardness figures are a measure of the grain size of the carbide, the highest hardness representing the finest grain size of less than 1 μm. By comparison the cobalt content has a relatively minor influence on the rate of attrition wear. *Figure 6.43* shows the small difference in rates of wear of carbide tools with 5.5–20% cobalt, all of the same grain size with large differences in hardness. Consistent performance under these conditions depends on the ability of the manufacturer to produce fine-grained alloys with close control of the grain size.

Since the metal flow around the tool edge tends to become more laminar as the cutting speed increases, the rate of wear by attrition may increase as the cutting speed falls. *Figure 6.44* shows a family of curves for the flank wear rate when cutting cast iron under mainly attrition wear conditions and should be compared with *Figure 6.37* for diffusion wear. To improve tool life where attrition is dominant, attention should be paid to reducing vibration, increasing rigidity and providing adequate clearance angles on the tools.

Abrasive wear

Because of the high hardness of tungsten carbide, abrasive wear is much less likely to be a significant wear process with cemented carbides than with high speed steel. There is little positive evidence of abrasion except under conditions where very large amounts of abrasive material are present, as with sand on the surface of castings. It seems very unlikely that isolated small particles of hard carbide or of alumina in the work material can be effective in eroding the cobalt from between the carbide grains under conditions of seizure. The wear of tools used to cut chilled iron rolls, where much cementite and other carbides are present, may be by abrasion, but most of the carbides, even in alloy cast iron, are less hard than WC and detailed studies of the wear mechanism in this case have not been reported. To resist abrasive wear, a low percentage of cobalt in the cemented carbide is the most essential feature, and fine grain size also is beneficial.

Fracture

Erratic tool life is often caused by fracture before the tool is much worn. The importance of toughness in grade selection, and the improvement in this property which results from increased cobalt content or grain size have already been discussed. Considerable care is necessary in diagnosing the cause of fracture to decide on the correct remedial action.

It is rare for fracture on a part of the tool edge to occur while it is engaged in continuous cutting. More frequently the tool factures on starting the cut, particularly if the tool edge comes up against a shoulder, so that the full feed is engaged suddenly. Interrupted cutting and operations such as milling are particularly severe and may involve fracture due to mechanical fatigue. A frequent cause of fracture on the part of the edge not engaged in cutting is impact by the swarf curling back onto the edge or entangling the tool. This is particularly damaging if the depth of cut is uneven, as when turning large forgings or castings.

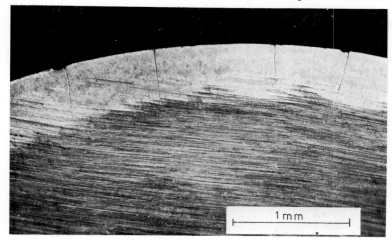

Figure 6.45 Thermal fatigue cracks in cemented carbide tool after interrupted cutting of steel

Fracture may be initiated also by deformation of the tool, followed by crack formation, the mechanical fracture being only the final step in tool failure. This case emphasises the importance of correct diagnosis, since the action to prevent failure in this case would include the use of a carbide with higher hardness, to prevent the initial plastic deformation, but less toughness. Prevention of fracture is rarely a problem which can be solved by changes in the carbide grade alone, and more often involves also the tool geometry and the cutting conditions.

Thermal fatigue

Where cutting is interrupted very frequently, as in milling, numerous short cracks are often observed in the tool, running at right angles to the cutting edge, *Figure 6.45*. These cracks are caused by the alternating expansion and contraction of the surface layers of the tool as they are heated during cutting, and cooled by conduction into the body of the tool during the intervals between cuts. The cracks are usually initiated at the hottest position on the rake face, some distance from the edge, then spread across the edge and down the flank. Carbide milling cutter teeth frequently show many such cracks after use, but they seem to make relatively little difference to the life of the tool in most cases. If cracks become very numerous, they may join and cause small fragments of the tool edge to break away. Also they may act as stress-raisers through which fracture can be initiated from other causes. Many carbide manufacturers have therefore selected the compositions and structures least sensitive to thermal fatigue as the basis for grades recommended for milling.

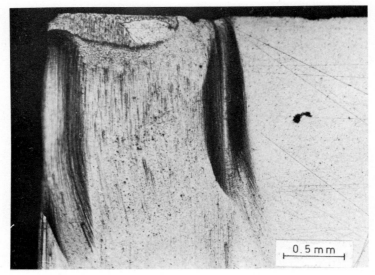

Figure 6.46   Sliding wear under edges of chip on rake face of carbide tool used for cutting steel

Wear under sliding conditions

Accelerated wear often occurs at those positions at the tool work interface where sliding occurs, as with high speed steel tools (see page 91). A pronounced example in the case of a WC–Co tool used to cut steel in air is shown in *Figure 6.46,* the deep grooves are at the positions where the edges of the swarf slid over the rake face of the tool.   The wear mechanism here probably involves reactions with the atmosphere, and is discussed further in Chapter 8 in relation to cutting lubricants.   It is of interest in showing that sliding may cause much more rapid wear than seizure under the same cutting conditions, and the elimination of seizure is, therefore, not a desirable objective in many cutting tool operations. The rate of wear in the sliding areas is mainly controlled by a chemical interaction and depends more on the composition of the tool material than on its hardness or other mechanical properties.

The mechanisms and processes described appear to be the main ways in which *carbide tools* are worn or change shape so that they no longer cut efficiently.   Many of the mechanisms are the same as with high speed steel tools and, in summing up, *Figure 6.19* can be referred to.   Mechanism *1* is not normally observed.   Mechanisms *2* and *3,* based on deformation and diffusion, are temperature dependent and come into play at high cutting speeds, limiting the rate of metal removal which can be achieved.   Mechanism *4,* attrition wear, is not temperature dependent, and is most destructive of the tools in the low cutting speed range, where high speed steels often give equal or superior performance.   Mechanism *5,* abrasion, is probably a minor cause of wear, of less significance

than on steel tools. Sliding wear processes, mechanism 6, may be important as with steel tools and occur at the same positions on the tool. In
addition, carbide cutting edges are more sensitive to failure by fracture,
and thermal fatigue may cause cracking which shortens tool life. The
correct diagnosis of factors controlling tool life in practical machining
operations is of major importance, both for the practical man in selection
of the optimum tool material, tool design or cutting conditions, and for
the research worker engaged in developing new tool alloys, tool shapes or
lubricants.

Tungsten-titanium-tantalum carbide alloys bonded with cobalt

The tungsten carbide-cobalt tools developed in the late 1920s proved very
successful for cutting cast iron and non-ferrous metals, at much higher
speeds than is possible with high speed steel tools, but they were less
successful for cutting steel. This was because of the cratering-diffusion
type of wear which caused the tools to fail rapidly at speeds not much
higher than those used with high speed steel. The success of tungsten
carbide when cutting cast iron encouraged research and development in
cemented carbides. Much effort was put into investigating carbides of
the other transition elements (*Table 6.2*).

The most successful of these have been titanium carbide (TiC), tantalum
carbide (TaC), and niobium carbide (NbC). All of these can be bonded
by nickel or cobalt, using a powder metallurgy technology similar to that
developed for the WC–Co alloys. In spite of all the effort put into their
development, none of these has so far been as successful as the WC–Co
alloys in combining the properties required for a wide range of engineering applications. In particular, the other cemented carbides lack the toughness which enables the tools based on WC to be used, not only for heavy
machining operations, but also for rock drills, coal picks and a wide range
of wear resistant applications. Two of these carbides, however, have come
to play a very important role as constituents in commercial cutting tool
alloys.

In the early 1930s tools based on TaC and bonded with nickel were
marketed in USA under the name of 'Ramet' and were used for cutting
steel because they were more resistant to cratering wear than the WC–Co
compositions. Alloys based on TaC never became a major class of tool
material, but they did indicate the direction for development of carbide
tool materials for cutting steel, i.e. the inclusion in the WC–Co alloys of
proportions of one or more of the cubic carbides – TiC, TaC or NbC.
In the early years of their development, tools made from these alloys
were often more porous than those of WC–Co, but as a result of the
research and development work by Dr P. Schwarzkopf[15] and many others,
the technology of production is now fully mastered, and their general
quality is very high. They are generally known as the *steel cutting
grades* of carbide, and all major manufacturers of cemented carbides make

two main classes of cutting tool materials – the WC–Co alloys (often called *the straight grades*) for cutting cast iron and non-ferrous metals, and the WC–TiC–TaC–Co alloys for cutting steels.

Table 6.5 Properties of some steel cutting grades of cemented carbide

Composition			Mean grain size	Hardness	Transverse rupture strength		Young's modulus		Specific gravity
Co%	TiC%	TaC%	μm	HV30	N mm^{-2}	tonf/in^2	N mm^{-2} X 10^3	tonf/in^2 X 10^3	
9	5	—	2.6	1 475	1 890	122	566	36.6	13.35
9	9	12	3.0	1 450	1 930	125	510	33.0	12.15
10	19	15	3.0	1 525	1 410	91.5	455	29.4	10.30
5	16	—	2.5	1 700	1 230	80	537	34.8	11.40

STRUCTURE AND PROPERTIES

The compositions and some properties of representative steel cutting grades of carbide are given in *Table 6.5*. Alloys containing from 4 to 60% TiC and up to 20% TaC by weight are commercially available, but those with more than about 20% TiC are made and sold in very small quantities for specialised applications. The proportions of cobalt and the grain size of the carbides are in the same range as for WC–Co alloys. Micro-examination shows that, in the structures of alloys containing up to about 25% TiC, two carbide phases are present instead of one, *Figure 6.47*. Both angular blue-grey grains of WC and rounded grains of the cubic carbides, which appear yellow-brown by comparison, are present. In alloys with more than 25% by weight of TiC, no WC grains can be detected, and, at lower TiC contents, it is obvious that the proportion of the cubic carbides is much greater than could be accounted for by the relatively small percentage by weight of TiC present. The explanation lies in the fact that WC can be taken into solid solution in TiC in large amounts (up to 70% by weight of WC can be dissolved in TiC) while TiC and TaC are both completely insoluble in WC. TiC and TaC have the same type of cubic structure and are completely soluble in one another. The rounded grains seen in the structure of these alloys are, therefore, grains of carbide with a cubic crystal structure, in which there are atoms of the metals Ti, Ta and W, and one carbon atom for each metal atom. The cubic phase in these alloys was first shown to be a solid solution in Germany, where the word *Mischkristalle* is the term for a *solid solution,* and those engaged in cemented carbide technology refer to it as the *mixed crystal* phase.

Table 6.5 shows that the hardness of the steel cutting grades is in the same range as that of the WC–Co alloys of the same cobalt content and

Figure 6.47 Structure of steel cutting grade of carbide containing WC, TiC, TaC and cobalt

grain size (*Table 6.3*). Increasing the Ti, Ta content reduces the transverse rupture strength, and practical experience in industry confirms that the toughness is reduced. For this reason the most popular alloys for cutting tools contain relatively small amounts of TiC/TaC. TiC is much cheaper than TaC, titanium being plentiful and tantalum being a rare metal, but TaC is considered to cause less reduction in toughness, and most present day alloys contain some TaC. The addition of the cubic carbides lowers the thermal conductivity, which is high for WC–Co alloys. The conductivity of an alloy with 15% TiC is approximately the same as that of high speed steel, and only about half of that for the corresponding WC–Co alloy.

PERFORMANCE OF WC– TiC–TaC–Co ALLOY TOOLS

The steel cutting grades of carbide have two major advantages. The hardness and compressive strength at high temperature are higher than those of WC–Co alloys (*Figure 6.5*). When cutting steel they are much more resistant to wear based on diffusion which causes cratering and rapid flank wear on WC–Co alloy tools. The steel cutting grades can be used at speeds often three times as high as the tungsten carbide 'straight grades', as is shown by comparing the machining chart in *Figure 6.48* for a tool containing 15% TiC with that in *Figure 6.36* for a WC–Co tool cutting the same steel. Rapid cratering occurred at 90 m min^{-1} (300 ft/min) at a feed rate of 0.25 mm (0.010 in) per rev feed on the WC–Co tool and at 270 m min^{-1} (900 ft/min) on the tool with 15% TiC. The higher

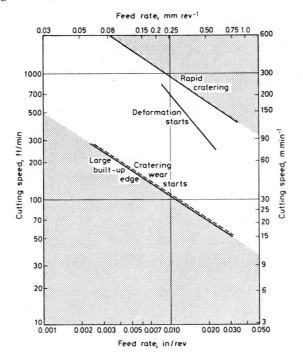

Figure 6.48 *'Machining chart' for steel cutting grade of carbide (Wimet XL3) cutting 0.4% C steel with Hardness 200 HV*[35]

the proportion of cubic carbides in the structure, the higher is the permissible speed when cutting steel, but even 5% of TiC has a very large effect. Not only cratering, but also flank wear is reduced when cutting steel in the higher cutting speed range. Metal removal rate when cutting steel with these alloys is limited not by the ability of the tool to resist cratering but by its ability to resist deformation or by too rapid flank wear. The choice of grade for any application depends on achieving the correct balance in the tool material between toughness and resistance to flank wear and deformation.

The process by which the cubic carbides function to reduce wear rate can be demonstrated by metallography. *Figure 6.49a* is the crater on the rake face of a WC–Co alloy tool after cutting a medium carbon steel at 61 m min⁻¹ (200 ft/min) and 0.5 mm (0.020 in) per rev feed rate for 30 s, while *Figure 6.49b* is the rake surface of a tool with 15% TiC used for cutting under the same conditions for 2.5 min. Microexamination of the worn surface shows that the cubic 'mixed crystal' grains are worn at a very much slower rate than the WC grains, and are left protruding from the surface to a height as great as 4 micrometres. *Figure 6.50* is a scanning electron microscope picture of such a worn

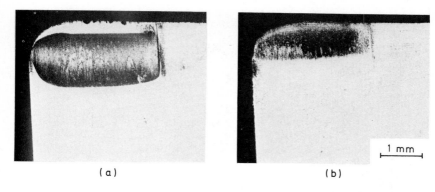

(a)   (b)

Figure 6.49   (a) Crater on rake face of WC-Co tool after cutting steel at high speed
(b) Crater on rake face of WC-TiC-Co tool after cutting under the same conditions
for 5 times as long

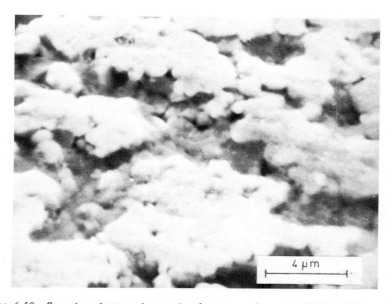

Figure 6.50   Scanning electronmicrograph of crater surface of WC-TiC-TaC-Co tool

surface after cleaning by removal of the adhering steel in acid.   *Figure*
6.51 is an optical microscope picture of a worn crater surface.   The surface of the worn WC grains is characteristic of chemical attack (diffusion)
rather than mechanical abrasion.   The 'mixed crystal' grains also are worn
by diffusion but at a very much slower rate at the same temperature.
If cutting is continued, the mixed crystal grains may be undermined and
carried away bodily in the under surface of the chips.   Cratering is therefore observed on tools of the steel cutting grades of carbide after cutting
steel at high speed, but at a much slower rate than on **WC**-Co alloy
tools.

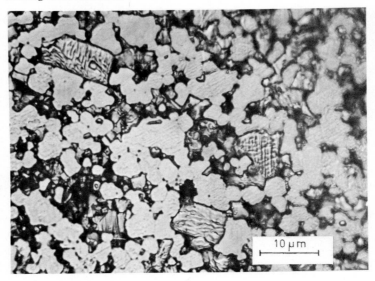

Figure 6.51   Photomicrograph of crater surface of WC-TiC-Co tool showing 'etching', diffusion attack on WC grains[25]

The introduction of the steel cutting grades of carbide completed the revolution in machine shop practice which was initiated by the WC–Co alloys.   Their importance lies in the fact that such a very large proportion of the total activity of metal cutting is concerned with the machining of steel.   The basic development of the steel cutting carbides had been accomplished by the late 1930s, and they played a large role in production on both sides in the Second World War.   Productivity of machine tools was increased often by a factor of several times and a complete revolution in machine tool design was required of importance equal to that which followed the introduction of high speed steel 40 years earlier.   Development of the steel cutting grades of carbide has continued, and tool tips of very high quality and uniform structure are available giving very consistent performance.   Minor modifications to composition and structure have been introduced to give improvements for particular applications such as milling, but until very recently no major new line of development has achieved outstanding success.

Development of the steel cutting grades was the result of intelligent experimentation and empirical testing.   Explanation of their performance, in terms of a deeper understanding of the wear processes during cutting, has followed much later and is still continuing.   Details of the mechanism of diffusion wear are still a subject of research, but knowledge of the main features of this wear process can form one of the useful guide lines for those involved in tool development and application.   The very large improvement in tool performance achieved by the introduction of the cubic carbides results from the low rate at which they diffuse into (dissolve in) the hot steel moving over the tool surface. The improvement is specific to *the cutting of steel* at *high rates of metal removal.* If either the temperature at the tool-work interface is too low, or work material other than steel or iron are machine, the advantages gained by using the steel cutting grades disappear and tool performance is often better when using one of the WC–Co grades. When machining steel at speeds and feed rates in the region of the built-up edge line (*Figures 6.36* and *6.48*) or below, the usual wear mechanism is attrition, and the steel cutting grades are more rapidly worn by attrition than are the WC–Co grades, the cubic carbides being more readily broken up than WC. For this reason WC–Co alloys are commonly used when machining steel where high speeds are impossible, for example on multi-spindle automatic machines fed with small diameter bars.

Titanium is a high melting point metal and high temperatures are generated at the tool-work interface during cutting, but WC diffuses less rapidly than TiC from the tool surface into titanium. The difference is not so great as the reverse effect in steel cutting, but as a consequence, the steel cutting grades have less resistance to diffusion wear when cutting titanium and its alloys,[29] and the WC–Co alloys are always used for this purpose.

The cubic carbides also have no advantage in terms of wear resistance at those parts of the tool where sliding takes place at the interface. Comparison of the performance of the cemented tungsten carbides with and without the addition of the cubic carbides demonstrates very well that wear rate is very dependent on the type of wear process or mechanism involved. Thus **wear resistance is not a unique property of a tool material which can be determined by one simple laboratory test, or correlated with one simple property such as hardness.** Correct diagnosis of the controlling wear mechanism for a particular operation can often be the starting point in selecting the optimum tool material.

Techniques of using cemented carbides for cutting

Steel tools are normally made in one piece, but from their earliest days cemented carbides were made in the form of tool tips – small inserts which were brazed into a seat formed in a steel shank. Except for very small tools it is uneconomic to make the whole tool of cemented carbide (which typically costs 20 times as much as high speed steel, weight for weight) and there is normally no advantage in terms of performance. A brazed carbide tool is used until it no longer cuts efficiently and is then reground. Most grinding of carbide tools requires the use of bonded diamond grinding wheels, more skill is required and the cost is greater than for grinding high speed steel tools.

Because cemented carbides are less tough than high speed steel, the tool edge must be more robust. With steel tools the rake angle may be as high as 30°, so that the tool is a wedge with an included angle as small as 50° (*Figure 2.2*). The rake angle of carbide tools is seldom greater than +10° and is often negative (*Figure 2.2d*). A tool with an

Figure 6.52 'Throwaway' tool tips of cemented carbide

included angle of 90° set at a negative rake angle of −5° is particularly useful for conditions of severe interruption of cut, or when machining castings or forgings with rough surfaces. The common use of tools with low positive rake, or negative rake angles made possible the concept of *throw-away tool tips.* These are small tablets of carbide, which are usually square or triangular in plan, though other shapes may be used (*Figure 6.52*). Each tip has a number of cutting edges – positive rake, triangular tips have three cutting edges, while on a negative rake square tip, 8 cutting edges can be used. In use the tips are mechanically clamped to the tool shank using one of a number of different methods. When a cutting edge is worn so that the tool no longer cuts efficiently, the tip is unclamped and rotated to an unused corner, and this is repeated until all the cutting edges are worn and then the tip is discarded. rather than reground. Very little of the tool material is worn off when the tool tip is thrown away, but, in spite of this, in many operations very considerable economies are achieved compared with the use of brazed tools which are reground. The high cost of regrinding is eliminated, the time required to change the tool is greatly reduced, and the cost of brazing is avoided. There is, however, only a limited range of operations for which throw-away tips are suitable, and brazed tools continue to be produced.

The powder metallurgy process is well adapted to mass production of the tool tips. These are made by pressing on automatic presses, followed by a sintering operation, which produces tips of high dimensional accuracy, so that little grinding is required. The introduction of throw-away tooling has made possible important new lines of development in tool shapes and materials. The tool edges can be given considerable protection against chipping caused by mechanical impact by slightly rounding the sharpened edge. Many manufacturers supply *edge-honed* tips, on the cutting edges of which there is a small radius, usually about 25 to 50 μm (0.001–0.002 in). Such tips can give much more consistent performance, without the rate of wear being increased.   Tips are also made with grooves formed in the rake face close to the edge which curl the chips as they are produced into shapes which can be readily cleared from around the tool – a problem encountered when cutting at high speeds.

The most important aspect of throw-away tool tips is the possibility which they open up of new lines of development in tool materials.   The new paths of tool material evolution are made possible because (1) a brazing operation is avoided and (2) only the surface material near the edge need be highly resistant to wear, while the body of the tool tip can be made of a tougher material.

Continued development of tool materials

While high speed steel and cemented carbides, based on WC, account for a very large proportion of all cutting tools used in engineering, research and development continue to produce new tool materials.   Some of the more interesting recent developments are now considered.

TiC-BASED TOOLS

Those steel cutting grades of carbide which contain large percentages of TiC are difficult to braze and were unpopular for this reason when brazed tools were the norm.   With throw-away tool tips, alloys with higher proportions of TiC could more readily be used and consideration was given to tools based on TiC instead of WC, because of its resistance to diffusion wear in steel cutting.

Of all the cubic carbides, TiC has the most obvious potential.   Titanium is a plentiful element in the earth's crust, the oxide TiO_2 is commercially available in purified form, and TiC can be readily made directly from the oxide at temperatures about 2 000 °C.   Cemented TiC alloys can be made by a powder metallurgy process differing only in detail from that used for the production of the WC–based alloys.   The most useful bonding metal has been nickel and usually the alloys for cutting contain 10–20% Ni.   The difficult problems of producing a fine-grained alloy of consistently high quality, free from porosity, have largely been overcome.   The addition of about 10% molybdenum carbide (Mo_2C) is often made to facilitate sintering to a good quality.   Many commercial suppliers now include a TiC-based grade in their catalogues.[30]

The TiC-based carbides have hardness in the same range as that of the conventional cemented carbides.   Experience, both in laboratory tests and in many industrial applications, shows that the TiC-based tools have lower rates of wear when cutting steel at high speed, and can be used to higher cutting speeds than the conventional steel cutting grades of carbide.   The advantages in terms of increased metal removal rate, however, in changing from conventional steel cutting grades to TiC-based carbides is not so great as that which was achieved when changing from WC–Co alloys to the steel cutting grades.   The TiC-based tools appear to lack the reliability and consistency of performance of the conventional cemented carbides and operators do not have confidence in their ability to apply them to a wide range of applications without trouble. In spite of their clear advantage in terms of lower wear rate, they have not yet accounted for more than a very small percentage of the tools in commercial use. There is probably a lack of toughness in the alloys so far made. This aspect of tool materials needs further study, because, with the relative shortage of tungsten supplies, a substitute for the WC-based alloys will eventually be required, and those based on TiC seem the most likely candidates so far.

LAMINATED TOOLS

To take advantage of the new degree of freedom afforded by throw-away tool tips, laminated tools have been made, in which the rake faces are coated with a thin layer, about 0.25 mm (0.010 in) thick, of a steel cutting grade of carbide, while the main body of the tip consists of a tough grade of WC–Co composition with high thermal conductivity. Production of such composite bodies by powder metallurgy techniques is possible using automatic presses. When using these laminated tips, reduced rates of wear are achieved compared with conventional cemented carbides, and increased cutting speeds are possible when cutting both steel and cast iron. One firm has marketed these laminated tools in the UK.[31]

COATED TOOLS

Further development of laminated tools has been superseded by the coating of conventional carbide tool tips with a very thin layer of a hard substance by chemical vapour deposition ('CVD' coatings).[32] TiC coatings were the first to be produced commercially, but tips with TiN (titanium nitride) coatings are also on the market, and further development of the coating process can be expected. The coatings are normally very thin — of the order of 5 μm (0.0002 in) and consist of a layer of pure carbide and/or nitride, metallurgically bonded to the cemented carbide tool tip.

To coat the tips they are heated to a temperature of the order of 1 000 °C in an atmosphere of hydrogen containing a small percentage of vapour of titanium tetrachloride ($TiCl_4$). The chloride decomposes at the tool surface to deposit a thin, continuous layer of titanium, which is carburised to TiC by diffusion of carbon from the carbide of the tool. Carburisation can also be achieved by including methane in the gaseous atmosphere. Similarly, the nitride TiN can be deposited by including NH_3 in the atmosphere, the nitride-coated tips being distinctive in having a golden colour. If the coatings are thin, of the order of 5 μm, they are found to be continuous, almost free from cracks and of very fine grain size. *Figure 6.53* shows a micro-section through a TiC layer, a typical

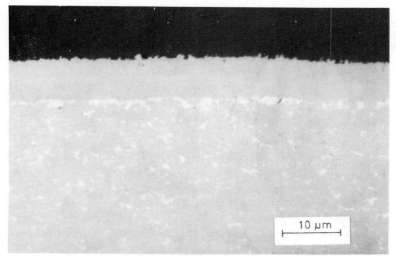

Figure 6.53   Section through TiC layer on coated cemented carbide tool tip

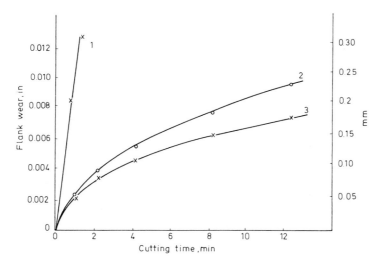

Figure 6.54   Tool wear vs time for coated and uncoated cemented carbide tool tips.   (1) uncoated steel cutting grade   (2) coated WC-Co alloy   (3) coated steel cutting grade

grain size of which, observed in electron microscope studies is 0.1 μm, an order of magnitude smaller than the carbide grains in the tool.

It is surprising that such very thin layers of pure carbide remain uncracked, strongly adherent and effective in wear resistance in practical situations. When cutting cast iron or steel at high speeds they have proved capable of increasing the life of normal throw-away cemented carbide tools by a factor of several times, when compared with the uncoated tips, or of increasing the rate of metal removal by operating at higher cutting speed, *Figure 6.54.* Both the WC–Co alloys and the steel cutting grades can be coated, the most suitable combination for successful application to a range of operations being determined by extensive industrial testing. They are largely successful in combining the advantages of the TiC-based tools in terms of metal removal rate, with the toughness and reliability of conventional cemented carbides when cutting steel and iron at high speeds. Operations involving severe interruptions of cut, however, may fracture and flake away the coating. The wear resistant qualities are specifically against the diffusion wear process in the high cutting speed range and they are less successful where wear is by attrition. They are, of course, unsuitable for any tool which must be reground.

The coating process has already achieved considerable sophistication, and further developments aim to strengthen adhesion, and to produce coatings by vapour deposition which are more wear resistant than TiC for specific applications. The vapour deposition of TiC on steel tools for forming and other operations actually preceded its use on carbide tools. There is some difficulty in reconciling the heat treatment of tool steels with the vapour deposition treatment. This problem may be overcome, but the application to steel tools for cutting operations is likely to be limited, because there are few steel tools which are not reground after service. Many new developments in coated tools can be expected.

CERAMIC TOOLS

Many other substances of high hardness and melting point have been investigated as potential cutting tool materials, and these have included the refractory oxides. The only one to be successfully applied in industrial cutting applications has been aluminium oxide, experiments with which started before the Second World War. Throw-away tool tips consisting basically of Al_2O_3 (alumina) have been available commercially for more than 20 years, and have been used in many countries for machining steel and cast iron.[33]

The successful tool materials consist of fine-grained (less than 5 μm) Al_2O_3 of high relative density, i.e. containing less than 2% porosity. Several different methods have been used to make tool tips which combine these two essential structural features, including (1) pressing and sintering of individual tips by a process similar to that used for cemented carbides, and (2) hot pressing of large cylinders of alumina in graphite moulds, the tool tips being cut from the cylinders with diamond slitting wheels.

The possibility of 'cementing' alumina particles together with a metal bond, using a process similar to the bonding of carbide by cobalt, has been explored, but no satisfactory metal bond has been found. However, many additions, e.g. MgO and TiO, have been made to promote densification and retain fine grain size. One commercial tool material contains roughly 50% carbide and 50% alumina. The basic raw material, alumina, is cheap and plentiful, but the processing is expensive and the tool tips are therefore not cheap compared with cemented carbides.

The room temperature hardness of alumina tools is in the same range as that of the cemented carbides, 1550–1700 HV. Their major advantages are (1) retention of hardness and compressive strength to higher temperatures than with carbides and (2) much lower solubility in steel than any carbide – they are practically inert to steel up to its melting point. To offset these advantages, their toughness and strength in tension are much lower. Unfortunately, no test method exists by which the toughness of this class of materials can be compared quantitatively with that of carbides. The transverse rupture strength is an inadequate guide to toughness but values reported range from 390–780 N mm^{-2} (25–50 tonf/in^2), about one third that of cemented carbides. Alumina is non-metallic in character, with an ionic rather than a metallic bond, and, consequently, is an electrical insulator with poor thermal conductivity. It is a true ceramic, the pure alumina being white, translucent and looking like porcelain to which it is akin. Its lack of toughness is therefore not surprising and it is unexpected to find that there are conditions where it can withstand the temperatures and stresses of cutting.

Alumina tools can be used to cut steel at speeds much higher than can be used with conventional cemented carbides or TiC-based alloys. Negative rake throw-away tool tips are nearly always used, and it is not difficult to demonstrate that cutting speeds of 600–750 m min^{-1} (2 000–2 500 ft/min) can be sustained, at a feed rate of 0.25 mm (0.010 in) per rev, for long periods without excessive wear when cutting cast iron and many steels. This is approximately three to four times the speeds normally used with cemented carbide tools, and represents an increase in metal removal rate as great as that achieved by high speed steel and by cemented carbides at their inception. However, in spite of the efforts of dedicated enthusiasts to introduce these tools on a large scale, the numbers in use are only a very small percentage of the carbide tool applications. Their main continuous usage is in cutting grey cast iron where a very good surface finish is required. On clutch facings and brakes for cars they are being used at speeds up to 600 m min^{-1} (2 000 ft/min) to give a surface finish good enough to eliminate a subsequent grinding operation.

It is difficult to find out why alumina tools have not come into more widespread use during the last 20 years. There are several cases where concentrated efforts have resulted in ceramic tools being used on numerous operations in individual factories for cutting both steel and cast iron. In the long run, however, their use has been discontinued except for isolated operations, mainly on cast iron. The most likely explanation is that inadequate toughness leads to unreliable performance under machine shop conditions, and that continuous attention to every detail of the operations is required to prevent premature tool failure. The interruption to production and damage to components caused by occasional failure, could more than off-set the economies gained by peak performance. The enormous economies which resulted from the introduction of high speed steel and cemented carbides naturally whet the appetite for a further round of speed increases on the same scale. However, there seems to be less potential advantage to be gained by increasing cutting speeds beyond those achieved by cemented carbides, for two reasons. The first is that, when cutting speeds are high, e.g. 200 m min^{-1} (600 ft/min), the time required to load and unload the work in the machine may be quite a large proportion of the cycle time, and further increases in cutting speed do not achieve a proportional reduction in total machining time. The second reason is that, at very high speeds, certain difficulties become accentuated – for example the chips coming off at very high speed are difficult to clear and may be a hazard to the operator. The chips are less of a problem when cutting cast iron because they do not form a continuous ribbon, and this is one reason for the successful application of ceramic tools to cast iron.

It seems doubtful whether alumina tools will come into widespread use or replace cemented carbides except in a small number of applications. The future line of development of tool materials may be towards greater consistency of performance, toughness and reliability rather than towards cutting speeds greatly exceeding those already achieved by coated carbide tips. There is, however, the possibility of making use of the wear-resistant qualities of alumina by coating carbide tool tips with thin layers of alumina, rather than TiC, and tools of this type are already being produced in several countries.

DIAMOND

The hardest of all materials, diamond, is used as a cutting tool, although its high cost restricts use to operations where other tool materials can not perform effectively. The hardness of diamond (approx. 6 000 HV at room temperature) is much higher than that of carbides or oxides and, under conditions of abrasive wear, it is outstanding, for example in the form of dies for drawing fine wire. Fine diamond powder is used for lapping, and wheels for grinding carbides and other hard materials are made by bonding diamond powder in a matrix of polymer, metal or ceramic. Large natural, industrial diamonds are in use as single point tools for cutting in specialist fields. They are lapped to the required tool shape, mounted in tool holders. Three examples of their use are (1) Finish-machining copper commutators – removing a small thickness of copper at fine feed and very high speed to produce a very good surface finish and very accurate dimensional control (2) Production of light alloy optical parts with good surface finish and dimensional accuracy (3) The finish-machining of aluminium alloy pistons with high silicon contents, where the large silicon crystals cause rapid wear on carbide tools. Diamond tools are also used for machining ceramics and partially sintered carbides. They are not usually recommended for cutting steel at high speed, because the diamond is worn by interaction with iron at high temperature. Diamonds are deficient in toughness – sharp edges are easily chipped and fractured, and this limits the range of operations in which they can be used.

The development of synthetic diamond manufacture has opened up another path of development. It is possible to make tool tips with a body of cemented carbide and a layer of synthetic diamond particles approximately 0.5–1 mm thick, sintered together and bonded onto the rake face of the tip. High density in the layer and strong bonding to the carbide are obtained by use of the same type of ultra-high pressure equipment as is employed for synthetic diamond manufacture. Diamond compact tools, made in this way, are more robust than the shaped natural diamond. The fine-grained sintered diamond is less readily fractured than the large, natural single crystals.[34]

The same method can be used for manufacture of tools with surface layers of cubic boron nitride (BN).[34] This is a 'man-made' material with a diamond type of structure and bonding, and having only slightly lower hardness than diamond. Boron nitride, made under the trade name of 'Borazon'[(R)] by The General Electric Company in USA has the advantages of being not easily oxidised in air, and being stable and inert to steel or nickel alloys to much higher temperatures than is diamond. It is therefore a potential tool material for cutting those difficult alloys, but since the cost of such tools is very great (suggested to be of the order of 60 times that of a carbide tool), their use is likely to be restricted to vital areas in which other tool materials are completely inadequate. One such application is the machining of creep-resistant alloys based on nickel, where tool life using carbide tools is extremely short. It seems clear that there are specialist fields for which such tools will prove economical.

General survey

At the beginning of this chapter the evolutionary character of the development of tool materials was emphasised. The evolutionary progress is demonstrated by stating the speeds commonly used industrially for cutting steel with different classes of tool material:

Tool Material	Cutting speed	
	m min^{-1}	ft/min
carbon steel	5	16
high speed steel	30	100
cemented carbide	150	500
ceramic	600	2000

This is dramatic but suggests much too simple a development. One might suppose that a more efficient tool material would quickly eliminate a less efficient one. However, because of the enormous variety of metal cutting conditions a range of tool materials remains and will continue to be used. It has been an objective of this chapter to offer some explanation for the continuing demand for a variety of tool materials.

Tool life is determined by a number of different mechanisms which change the shape of the tool edge until it no longer cuts efficiently. There is no simple relationship between tool performance and some property of the tool material such as hardness or wear resistance. 'Wear resistance' is not a unique property of a tool material, but is a complex interaction between tool and work materials very dependent on the cutting conditions. The subject of tool performance is bound to be complex, but some of the ways in which the behaviour of the two major classes of tool material are related to their structure, properties and composition have been described, with the objective of giving a guide to the selection of tool materials and their further development.

References

1. 'Tool Steel Feature', *Iron & Steel,* Special Issue, (1968)
2. BROOKES, K.J.A., *World Directory and Handbook of Hard Metals,* Engineers Digest (1975)
3. OSBORN, F.M., *The Story of the Mushets,* Nelson (1952)
4. TAYLOR, F.W., *Trans. A.S.M.E.,* **28,** 31 (1907)
5. MUKHERJEE, T., *I.S.I. Publication,* **126,** 80 (1970)
6. LOLADZE, T.N., *Industrie Anzeiger,* No.62, 991 (1959)
7. British Standard 4659: 1971
8. KIRK, F.A., CHILD, H.C., LOWE, E.M. and WILLIAMS, T.J., *I.S.I. Publication,* **126,** 67 (1970)
9. WEAVER, C., Unpublished work
10. TRENT, E.M., *Proc. Int. Conf. M.T.D.R., Manchester, 1967,* 629 (1968)
11. WRIGHT, P.K. and TRENT, E.M., *Metals Technology,* **1,** 13 (1974)
12. OPITZ, H. and KÖNIG, W., *I.S.I. Publication,* **126,** 6 (1970)
13. 'Ed. Supplement', *A.S.M. Handbook* (1948), also *Metal Progress,* 15 July, **141,** (1954)
14. DAWIHL, W., *Handbook of Hard Metals,* H.M.S.O. London (1955). (English translation)
15. SCHWARZKOPF, P. and KIEFFER, R., *Cemented Carbides,* Macmillan, New York (1999)
16. GOLDSCHMIDT, H.J., *Interstitial Alloys,* Butterworth (1967)
17. SCHWARZKOPF, P. and KIEFFER, R., *Refractory Hard Metals,* Macmillan New York (1960)
18. ATKINS, A.G. and TABOR, D., *Proc. Roy. Soc.* (A), **292,** 441 (1966)
19. MIYOSHI, A. and HARA, A., 'Powder & Powder Metallurgy', *J. Japan Soc.,* **12,** 2 (1965)
20. EXNER, H.D. and GURLAND, J., *Powder Metallurgy,* **13,** 5, 13 (1970)
21. FEATHERBY, M., *Ph.D. Thesis,* University of Birmingham (1968)
22. BURBACH, J., *Sonderdruck aus Technische Mitteillungen Krupp,* **26,** S 71-80 (1968)
23. 'Classification of carbides according to use', *I.S.O.,* 513 (1975)
24. TRENT, E.M., *I.S.I. Report,* No.94, 11 (1967)
25. TRENT, E.M., *I.S.I. Report,* No.94, 77 (1967)
26. TRENT, E.M., *I.S.I. Report,* No.94, 179 (1967)
27. TRENT, E.M., *J. Birmingham Met. Soc.,* **40,** 1, 1 (1960)
28. TRENT, E.M., *Proc. Inst. Mech. Eng.,* **166,** 64 (1952)
29. FREEMAN, R.M., *Ph.D. Thesis,* University of Birmingham (1975)
30. MAYER, J.E., MOSKOWITZ, D. and HUMENIK, M., *I.S.I. Publication,* **126,** 143
31. British Patent 1 042 711
32. SCHINTLMEISTER, W., and PACHER, O., *J. Vac. Sci. Technol.,* **12,** 4, 743 (1975)
33. KING, A.G. and WHEILDON, W.M., *Ceramics in Machining Processes,* Academic Press, New York (1960)
34. HIBBS, L.E. Jr. and WENTORF, R.H. Jr., *8th Plansee Seminar,* Paper No.42 (1974)
35. TRENT, E.M., *I.P.E.J.,* **38,** 105 (1959)

7
MACHINABILITY

The term *machinability,* which is used in innumerable books, papers and discussions, may be taken to imply that there is a property or quality of a material which can be clearly defined and measured as an indication of the ease or difficulty with which it can be machined. In fact there is no clear cut unambiguous meaning to this term. To the active practitioner in machining engaged in a particular set of operations, the meaning of the term is clear, and for him, machinability of a work material can often be measured in terms of the numbers of components produced per hour, the cost of machining the component, or the quality of the finish on a critical surface.

Problems arise because there are so many practitioners carrying out such a variety of operations, with different criteria of machinability. A material may have good machinability by one criterion, but poor machinability by another, or when a different type of operation is being carried out, or when conditions of cutting or the tool material are changed.

To deal with this complex situation, the approach adopted in this chapter is to discuss the behaviour of a number of the main classes of metals and alloys during machining, and to offer explanations of this behaviour in terms of their composition, structure, heat treatment and properties. The machinability of a material may be assessed by one or more of the following criteria:

1. *Tool life.* The amount of material removed by a tool, under standardised cutting conditions, before the tool performance becomes unacceptable or the tool is worn by a standard amount.
2. *Limiting rate of metal removal.* The maximum rate at which the material can be machined for a standard short tool life.
3. *Cutting forces.* The forces acting on the tool (measured by dynamometer, under specified conditions) or the power consumption.
4. *Surface finish.* The surface finish achieved under specified cutting conditions.
5. *Chip shape.* The chip shape as it influences the clearance of the chips from around the tool, under standardised cutting conditions.

As far as possible, all these criteria of machinability are taken into consideration when discussing each class of work material. The metals and alloys with the best machinability are discussed first.

Magnesium

Of the metals in common engineering use, magnesium is the easiest to
machine — the 'best buy' for machinability, scoring top marks by almost
all the criteria.   Rates of tool wear are very low because magnesium
does not alloy with steel, and the metal and its alloys have a low melting point (m.p. of Mg is 650 °C), so that temperatures at the tool-work
interface are low even at very high cutting speed and feed rate.   Turning speeds may be up to 1 350 m min^{-1} (4 000 ft/min) in roughing cuts,
and finishing cuts can be faster with good tool life.   Magnesium alloys
behave, in this respect, much like the pure metal.
    The tool forces when cutting magnesium are very low compared with
those when cutting other pure metals, and they remain almost constant
over a very wide range of cutting speeds,[1]   *Figure 7.1.*   Both the cutting force (F_c) and the feed force (F_f) are low and the power consumption is considerably lower than that when cutting other metals under

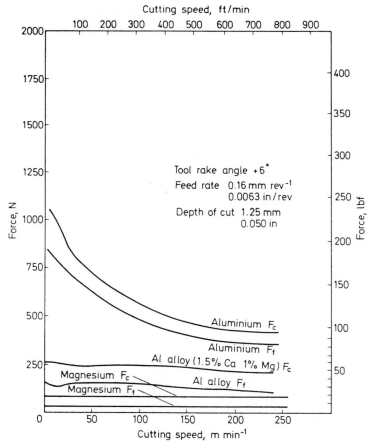

Figure 7.1   Tool forces vs *cutting speed — magnesium and aluminium.   (From data
by Williams, Smart and Milner[1])*

the same conditions. The low tool forces are associated with the low shear yield strength of magnesium, and, more important, with the small area of contact with the rake face of the tool over a wide range of cutting speeds and rake angles. This ensures that the shear plane angle is high and the chips thin — only slightly thicker than the feed.

Steel or carbide tools can be used and the surface finish produced is good at low and high cutting speeds. The chips formed are deeply segmented and easily broken into short lengths, so that chip disposal is not difficult even when cutting at very high speed. The hexagonal structure of magnesium is probably mainly responsible for the low ductility which leads to the fragile segmented chips and the short contact length on the tool rake face. The worst feature of the machinability of magnesium is the ease with which fine swarf can be ignited and the fire risk which this involves.

Aluminium

Alloys of aluminium in general also rate highly in the machinability table by most of the criteria. As with magnesium, the melting points of aluminium (659 °C) and its alloys are low and the temperatures generated during cutting are never high enough to be damaging to the heat-treated structures of high speed steel tools. Good tool life can be attained when cutting most aluminium alloys up to speeds of 600 m min^{-1} (2 000 ft/min) when using carbide tools and 300 m min^{-1} (1 000 ft/min) with tools of high speed steel. Speeds as high as 4 500 m min^{-1} (15 000 ft/min) have been used for special purposes.

Wear on the tool takes the form of flank wear, but no detailed study of the wear mechanism has been reported. High tool wear rates become a serious problem with only a few aluminium alloys. In aluminium–silicon castings containing 17–23% Si, where the silicon content is above the eutectic composition, the structures contain large grains of silicon, up to 70 μm across, in addition to the finely dispersed silicon of the eutectic structure. The large silicon crystals greatly increase the wear rate, even when using carbide tools.[2] The eutectic alloys, containing 11–14% Si can be machined at 300–450 m min^{-1} (1 000–1 500 ft/min) with good carbide tool life, but the presence of large silicon grains may reduce the permissible speed to only 100 m min^{-1} (300 ft/min). The drastic effect of large silicon particles is the result of the high stress and temperature which these impose on the cutting edge. *Figure 7.2* shows a section through the worn cutting edge of a carbide tool after cutting a 19% Si alloy. The layer of silicon attached to the worn surface demonstrates the action of the large silicon crystals which cause an attrition type of wear. The silicon particles have a high melting point (1 420 °C) and high hardness (> 400 HV). This action demonstrates that the wear of tools depends not only on the phases present in the work material, but also on their size and distribution. Fine silicon particles in the eutectic alloy can pass the cutting edge without severe damage

Figure 7.2   Section through cutting edge of cemented carbide tool used to cut 19% Si-Al alloy at 122 m min⁻¹ (400 ft/min)[3]

to the tool. The machining of high silicon alloys is one of the applications for diamond tools and for the recently developed diamond-coated tools.

In general, tool forces when cutting aluminium *alloys* are low, and tend to decrease slightly as the cutting speed is raised, *Figure 7.1.*   High forces occur, however, when cutting commercially pure aluminium particularly at low speeds. In this respect aluminium behaves differently from magnesium, but in a similar way to many other pure metals. The area of contact on the rake face of the tool is very large, and, as explained in Chapter 4, this leads to a high feed force (F_f), low shear plane angle, and very thick chips, with consequent high cutting force (F_c) and high power consumption. The effect on pure aluminium of most alloying additions or of cold working, is to reduce the tool forces, particularly at low cutting speed. In general most aluminium alloys, both cast and wrought, are easier to machine than pure aluminium, in spite of its low shear strength.

A built-up edge is not present when cutting commercially pure aluminium, but the surface finish tends to be poor except at very high cutting speed. Most aluminium alloys have structures containing more than one phase, and with these a built-up edge is formed at low cutting speeds, *Figure 7.3*. At higher speeds, e.g. above 60–90 m min⁻¹ (200–300 ft/min), the built-up edge may not occur. Tool forces are low where a built-up edge is present, and the chip is thin, but the surface finish tends to be poor.

The main machinability problems with aluminium are in controlling the chips. Extensive plastic deformation before fracture occurs more readily with aluminium, which has a face-centred cubic structure, than with the hexagonal magnesium. When cutting aluminium and some of its alloys, the chips are continuous, rather thick, strong and not readily broken. The actual form of the swarf varies greatly, but it may entangle the tooling and require interruption of operation to clear the chips.

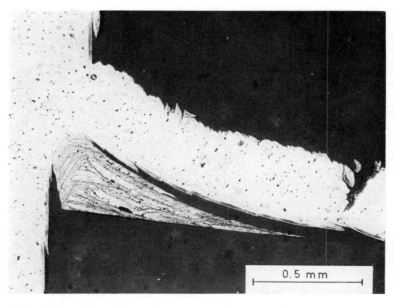

0.5 mm

Figure 7.3   Built-up edge when cutting Dural at 38 m min⁻¹ (125 ft/min)

In drills, taps and cutters of many types, it may clog the flutes or spaces between the teeth, so that modified designs of tools are often required for cutting aluminium. The cutting action can often be improved by modifications to rake and approach angles, or the introduction of chip breakers or curlers which deflect the chips into a tight spiral. Another approach is to modify the composition of the alloys to produce chips which are fragmented or more easily broken. The standard aluminium specifications now include 'free machining' alloys containing additions of lead, lead and bismuth, or tin and antimony in proportions up to about 0.5%. How these additions function is not certain, but the chips are more readily broken into small segments. These low melting point metals do not go into solid solution in aluminium, and are present in the structure as dispersed fine globules. They may act to reduce the ductility of the aluminium as it passes through the shear plane to form the chip. The main purpose of 'free-cutting' additives in aluminium and its alloys is improvement in the chip form rather than better tool life or an increase in the metal removal rate.

The excellent machinability of aluminium alloys in general makes them ideal work materials to be shaped in automated machine tools. Completely automatic production of certain classes of shapes can be introduced with confidence because long tool life and consistent performance can be guaranteed even at high rates of metal removal and when a great variety of operations is involved, such as turning, milling, drilling, tapping and reaming. Attempts to use the automated methods on other classes of work material have not been so successful.

Copper

Copper is another highly ductile metal with a face-centred cubic structure, like aluminium, but it has a higher melting point (1 083 °C). In general, copper-based alloys also have good machinability and for the same reason as with aluminium alloys.[4] Although the melting point is higher, it is not high enough for the temperatures generated by shear in the flow-zone to have a very serious effect on the life or performance of cutting tools. Both high speed steel and cemented carbide tools are employed. Good tool life is obtained, the wear on the tools being in the form of flank wear or cratering or both, but detailed studies of the wear mechanisms have not been reported. Even with carbon tool steels quite high cutting speeds are possible, and before the introduction of high speed steel, speeds as high as 100 m min[-1] (350 ft/min) were recommended when cutting brass.

The most important field of machining operations on copper-based alloys is in the mass production of electrical and other fittings using high speed automatic machines. These are mostly very high speed lathes, but fed with brass wire of relatively small diameter, so that the maximum cutting speed is limited to 140–220 m min[-1] (450–700 ft/min), although the tooling is capable of good performance at much higher speeds if required.

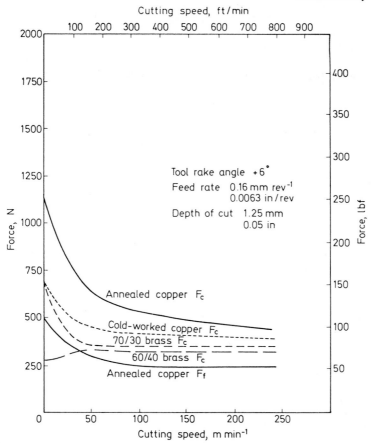

Figure 7.4   Tool forces vs cutting speed — copper and brass.   (From data by Williams, Smart and Milner[1])

The tool forces when cutting pure copper are very high, particularly at low cutting speed, *Figure 7.4,* and as with aluminium, this is essentially due to the large contact area on the rake face, resulting in a small shear plane angle and thick chips.   For this reason, high conductivity copper is regarded as one of the most difficult materials to machine. In drilling deep holes, for example, the forces are often high enough to fracture the drill.   Additional problems in the machining of pure copper are poor surface finish, particularly at low speeds, and the high strength of the tangled coils of continuous chips which are difficult to clear.

The machining qualities of copper are somewhat better after cold working, but are greatly improved by alloying. *Figure 7.4* shows the reduction in tool forces as a result of cold working which reduced the contact area, giving a larger shear plane angle and a thinner chip. The tool forces are lower for the 70/30 brass, which is single-phased, but there is a greater reduction when cutting the two-phased 60/40 brasses, the forces being low over the whole speed range, with thin chips and small areas of contact on the tool rake face. The forces are lowest in alloys of high zinc content where the proportion of the β phase is greater. The low tool forces and power consumption with the α–β brasses, together with low rates of tool wear, are a major reason for classifying them as of high machinability.

The chips from brass, however, are continuous, and to improve both chip disposal and surface finish, additives have been made to produce the 'free-machining brasses'. The usual additive is lead in the proportion of 2–3% by weight. This is soluble in molten brass but is rejected during solidification, precipitating particles usually between 1 and 10 micrometres (microns) in diameter, which should be uniformly dispersed to achieve good machinability. These additives greatly reduce tool forces, *Figure 7.5*, which become almost independent of cutting speed. Thin chips are produced, not much thicker than the feed, which are fragmented into very short

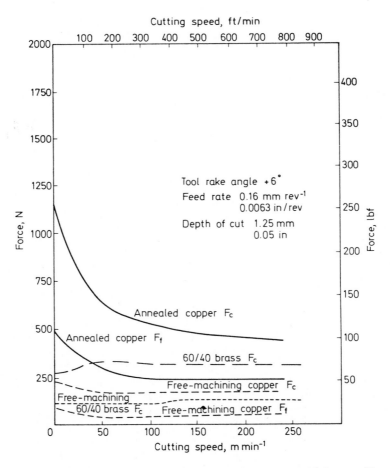

Figure 7.5   Tool forces vs *cutting speed – free cutting copper and brass.   (From data by Williams, Smart and Milner[1])*

lengths and are readily disposable, while the tool wear rate also is reduced. Free machining brass can be cut for long periods on automatic machines without requiring shut-down to replace tools or to clear swarf. Many small parts are economically made from free-machining brass because of the low machining costs and in spite of the high price of copper. The most probable reason for the success of lead in improving machinability, is that it reduces ductility, resulting in fracture or partial fracture on the shear plane. In addition to fragmenting the chip, this ensures that the compressive stress, which forces the chip against the tool, drops rapidly with distance from the cutting edge, so that the contact area is small.

Additions are made also to high conductivity copper to improve its machinability. Sulphur or tellurium are added to form plastic non-metallic inclusions – Cu_2S or Cu_2Te dispersed in the structure. Additions to copper have to be confined to those which 'do not appreciably reduce electrical conductivity or cause fracture during hot working. About 0.3% sulphur or 0.5% tellurium is usually added, reducing the electrical conductivity to about 98% of that of standard high conductivity copper. The effect is to reduce greatly the tool forces, particularly at low speeds, *Figure 7.5,* and to produce thin chips which curl and fracture readily. The surface finish is greatly improved.

All the two-phase alloys, including the free-cutting copper and the α–β brasses tend to form a built-up edge at low cutting speeds. This tends to disappear as the cutting speed is raised, e.g. over 30 m min^{-1} (100 ft/min), although a small built-up edge has been reported when cutting copper up to 600 m min^{-1} (2 000 ft/min).

Iron and steel

It is in the cutting of iron, steel and other high melting-point metals and alloys that the problems of machinability become of major importance in the economics of engineering production. With these higher melting-point metals the heat generated in cutting becomes a controlling factor, imposing constraints on the rate of metal removal and of tool performance, and hence on machining costs.

IRON

Commercially pure iron, like copper and aluminium, is in general a material of poor machinability. At room temperature the structure is body-centred cubic (α iron), transforming to face-centred cubic (γ iron) at just over 900 °C, and in both conditions it has relatively low shear strength but high ductility. The tool forces are higher than for copper (*Figure 4.5*), particularly at low cutting speed, but decrease rapidly as the speed is raised. The high tool forces are associated with a large contact area on the rake face of the tool, and thick chips. No built-up edge is observed at any speed, and quick-stops show a flow-zone about 25–50 μm (0.001–0.002 in) thick seized to the rake surface of the tool

(*Figures 3.10, 3.14*), which is the main heat source raising the temperature of the tool.

In discussing temperature distribution in cutting tools in Chapter 5, the problem is considered in relation to the cutting of a very low carbon steel, which is in effect a commercially pure iron. *Figures 5.6, 5.9* and *5.5* demonstrate the temperature pattern characteristic of tools used to cut this material, and this pattern is very important in relation to the machinability of iron and of steel. The upper limit to the rates of metal removal, using high speed steel and carbide tools, is determined by tool wear and deformation mechanisms controlled by these temperatures.

In general this same type of temperature pattern occurs in tools used to cut carbon and alloy steels, including austenitic stainless steels. A high temperature is generated on the rake face well back from the cutting edge, leaving a low temperature region near the edge. The temperature increases as cutting speed and feed rate are raised.

STEEL

The alloying elements added to iron to produce steel and to increase its strength (carbon, manganese, chromium etc.) influence both the stresses acting on the tool and the temperatures generated. As with copper and aluminium, the effect of alloying additions to iron is often to reduce the *tool forces* as compared with pure iron (*Figure 4.6*). The *stress* required to shear the metal on the shear plane to form the chip is greater when cutting steel, but the chip is thinner, the shear plane angle is larger and the area of the shear plane is much smaller than when cutting iron. We know that the *cutting force* is reduced by the addition of alloying elements, but how this affects the *compressive stress* acting on the rake face of the tool has not been investigated. The evidence reviewed in Chapter 4 (*Figures 4.8, 4.9*) demonstrates that stress is normally at a maximum near the edge. Numerical values of this stress when cutting steel, and how it is affected by alloying additions have not been determined.

The yield stress of steels is determined by both composition and heat treatment. When steels are heat treated to high strength and hardness, the compressive stress which they impose on tools during cutting becomes high enough to deform the cutting edge (*Figure 6.12*) and destroy the tool. Using high speed tools, the machining of steels with hardness higher than 300 HV becomes very difficult, even at low speeds where the tool is not greatly weakened by heat. Cemented carbide tools can be used to cut steels with higher hardness, but tool life becomes very short and permissible cutting speeds very low when the hardness exceeds 500 HV. *Figure 7.6* shows the speeds and feeds at which carbide tools were found to deform severely when cutting steels of different hardness, using a standard tool geometry.

To permit higher metal removal rates with most steels, it is therefore normal to heat treat to reduce the hardness to a minimum. The heat

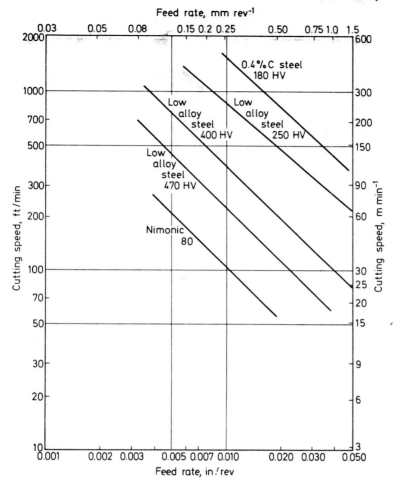

Figure 7.6   Conditions of deformation of cemented carbide tools when cutting steels of different hardness[5]

treatment often consists of annealing just below the transformation temperature (about 700 °C) to 'spheroidise' the cementite – the form in which it has least strengthening effect, *Figure 7.7*.   For some operations a coarse pearlite structure is preferred, a structure obtained by a full annealing treatment in which the steel is slowly cooled from above the transformation temperature.

Thus the limitation on rates of metal removal due to deformation of the tool can be related to the yield stress of the steel which can be determined by normal laboratory tests.   Of equal or greater importance however, is the influence of the alloying elements on the temperatures and temperature gradients generated in the tools.   These cannot at present be predicted from the mechanical properties because of the extreme conditions of strain and strain rate in the flow-zone, which is the heat source.

150

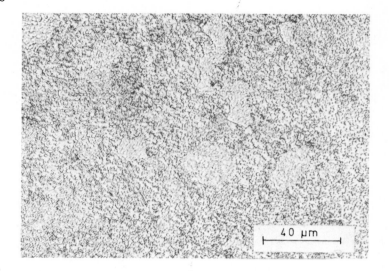

Figure 7.7   Structure of spheroidised steel

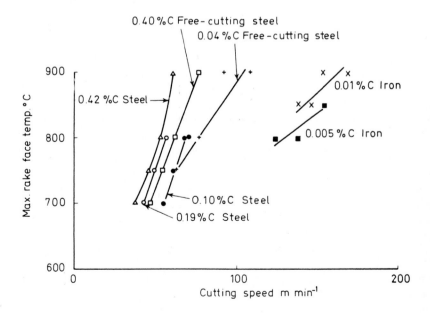

Figure 7.8   Maximum temperature on rake face vs *cutting speed when cutting steels of different carbon content.   (B.W. Dines[9])*

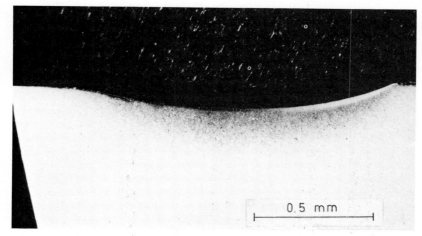

Figure 7.9   Temperature distribution in high speed steel used for cutting 0.4% C steel at 61 m min⁻¹ (200 ft/min).   (B.W. Dines[9])

This is discussed in Chapter 5.   Experimental study of the influence of alloying elements on the temperatures in the tools is a more effective way of investigating this aspect of machinability.   The few such studies carried out so far, show that the introduction into iron of strengthening alloying elements has two main effects on the temperatures in tools when cutting in the high speed range:

1. The same characteristic temperature gradient is maintained.
2. A lower cutting speed is required to generate the same temperature.

Figure 7.8[9] shows the maximum temperature observed on the rake face of tools used to cut steels with different carbon contents as a function of cutting speed, the feed rate being 0.25 mm (0.01 in) per rev.   The effect of carbon and other alloying additions is to accelerate the wear caused by temperature-dependent wear mechanisms both on high speed steel and on carbide tools, and thus to reduce tool life in the high speed cutting range.

Figure 7.9 shows a section through a high speed steel tool used to cut a 0.4% carbon steel at 61 m min⁻¹ (200 ft/min).[9]   A crater is being formed by shearing of the surface layers of the tool, but at a much lower cutting speed than when cutting the very low carbon steel (*Figure 5.3*), and at a lower temperature, because the higher carbon steel also imposes a higher shearing stress on the heated area.

Built-up edge

At lower cutting speeds, when machining steels containing more than about 0.08% carbon, and therefore having an appreciable amount of

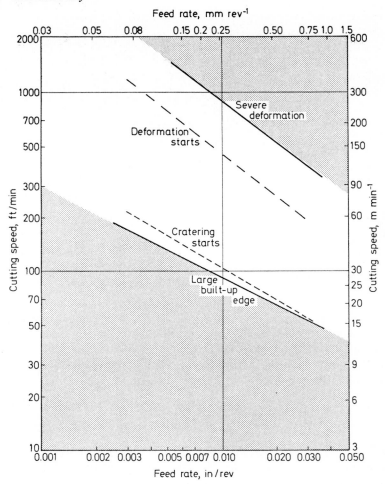

Figure 7.10 *'Machining chart' for steel cutting grade of carbide (Wimet XL3) used for cutting Ni-Cr-Mo steel — hardness 258 HV*

pearlite in the structure, a built-up edge is formed which has a major influence on all aspects of machinability (*Figure 3.15*).   As cutting speed is increased, a limit is reached above which a built-up edge is not formed. This limit is dependent also on the feed, and the conditions under which a built-up edge is formed are shown for two steels in the machining charts, *Figures 6.48* and *7.10,* for one standard tool geometry.   The influence of cutting speed on the built-up edge can be demonstrated simply by taking a facing cut on a lathe on a bar of steel rotating at a constant spindle speed.   If the cut is started from a small hole in the centre and the tool is fed outward, a built-up edge will be present at first, but the cutting speed continuously increases, and at a critical speed, the shape of the chip changes and the surface finish improves as the

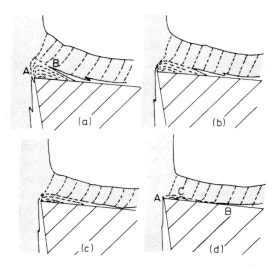

Figure 7.11   Transition from built-up edge to flow-zone with increasing cutting speed[10]

built-up edge disappears.   There are also conditions at very low rates of metal removal where a built-up edge is absent, but usually this is at speeds below 1 m min⁻¹ (3 ft/min).

The built-up edge, which is formed on both high speed steel and cemented carbide tools, consists of steel, greatly strengthened by extremely severe strain, the pearlite being much broken up and dispersed in the matrix.   Hardness as high as 600/700 HV has been measured on a built-up edge, which is considerably harder than steel wire of the highest tensile strength.   The built-up edge can therefore withstand the compressive and shearing stresses imposed by the cutting action.   When the cutting speed is raised, temperatures are generated at which the dispersed pearlite can no longer prevent recovery or recrystallisation and the structure is weakened until it can no longer withstand these stresses.   The built-up edge then collapses and is replaced by a flow-zone.   The transition can be pictured diagrammatically as in *Figure 7.11*.

In effect the built-up edge alters the geometry of the tool.   As *Figure 3.15* shows, it lifts the chip up off the rake face, so that the contact area to be sheared is much smaller than in the absence of a built-up edge.   This results in a large reduction in the forces acting on the tool as can be seen in *Figure 4.6*.   Power consumption is reduced and tool temperatures are relatively low.   Fragments of the built-up edge are constantly being broken away and replaced, *Figure 3.15,* but usually the fragments are relatively small.   Tool life may be rather erratic.   While the built-up edge may take over the functions of the tool cutting edge, so that the tools may be practically unworn for long periods of time, if intermittent contact with the tool edge occurs this leads to attrition wear.   As shown in Chapter 6, high speed steel tools are generally used under these conditions, because they often give much longer and more consistent tool life than cemented carbides.

Fragments of the built-up edge which break away on the newly formed work surface (*Figure 3.15*) leave this very rough, and better surface finish is usually produced by cutting at speeds and feeds above the built-up edge line on the machining charts, using carbide tools.

Free-cutting steels

The economic incentive to achieve higher rates of metal removal and longer tool life, has led to the development of the *free-cutting* range of steels, the main feature of which is high sulphur content, but which are improved for certain purposes by the addition of lead and sometimes tellurium. Typical free-cutting steel compositions are given in *Table 7.1*. The manganese content of these steels must be high enough to ensure that all the sulphur is present in the form of manganese sulphide (MnS), and steel makers pay attention not only to the amount but also to the distribution of this constituent in the steel structure. *Figure 7.12* shows the MnS in a typical free-cutting steel. Control of the MnS particle

Table 7.1   Typical compositions of free-cutting steels

Steel type	Percentage by weight			
	C	Mn	S	P
Low carbon	0.15	1.1	0.2 −0.3	0.07 max
	0.15	1.3	0.3 −0.6	0.07 max
'28 carbon'	0.28	1.3	0.12−0.2	0.06 max
'36 carbon'	0.36	1.2	0.12−0.2	0.06 max
'44 carbon'	0.44	1.2	0.12−0.2	0.06 max

shape, size and distribution is achieved during the steel making. It is influenced both by deoxidation of the molten steel before casting the ingots and by the hot rolling practice. Tellurium is present in steel as telluride particles which seem to perform a similar role to the sulphides. Tellurium has certain toxic properties which involve a hazard for the steel makers, and although evidence has been given of the improved machining qualities of tellurium steels, their use seems unlikely to become widespread. Lead is insoluble in molten steel, or nearly so, and good distribution of the lead is difficult to achieve. In order to disperse the lead in the form of fine particles throughout the steel, it is added in the form of shot to the steel as this is tapped from the ladle into the ingot moulds. When lead is added to high sulphur steels it is usually found attached to the MnS particles, often as a tail at each end.

The free-cutting steels are used extensively for mass production of parts on automatic machine tools, because they permit the use of higher cutting speeds, give longer tool life, give good surface finish, lower tool forces and power consumption, and produce chips which can be more readily handled.[6,7,8] Above all they are used because they can be relied

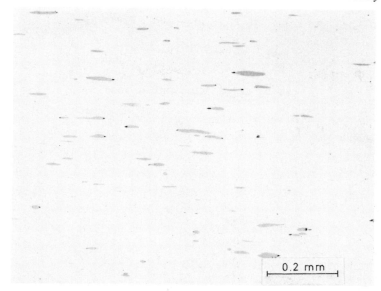

Figure 7.12   Structure of free-cutting steel showing MnS inclusions

on to perform more consistently than the non free-cutting steels in a wide variety of operations without interruption to the automatic machine cycle. In spite of extensive investigations into the mechanisms by which MnS acts to improve machinability, the explanations are still incomplete.

MnS particles dispersed in steel are plastically deformed when the steel is subjected to metal working processes.   Those in *Figure 7.12* had been elongated during hot rolling of the bar from the ingot.   In this respect they behave differently from the carbide (Fe_3C) plates in pearlite, which are fractured, or many small oxide particles which rigidly maintain their shape while the steel matrix flows around them.   There have been laboratory studies of the deformation of MnS particles, the extent of which depends on the amount of strain and the temperature.   Micro-examination of quick-stop sections shows that, on the shear plane, the sulphides are elongated in the direction of the shear plane, *Figure 7.13*. In the flow-zone adjacent to the tool surface, the elongation is very much greater.   *Figure 7.14* shows sulphide particles so drawn out in the flow-zone that their thickness is on the limits of resolution of the optical microscope, and some may be too thin to be seen, i.e. less than 0.1 μm.   In the flow-zone on the under surface of free-cutting steel chips a very high concentration of these thin ribbons of sulphide can be seen in scanning electron microscope pictures after etching in nitric acid, *Figure 7.15*.[9]   It is possible that, in this form, they may provide surfaces of easy flow where work done in shear is less than in the body of the metal.   Certainly the presence of the sulphide reduces the temperatures in high speed steel tools, *Figure 7.8*.[9]

156

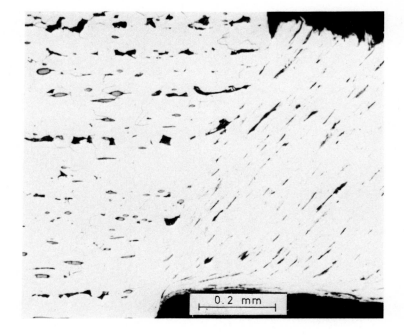

Figure 7.13   Deformation of MnS inclusions on the shear plane when cutting free-cutting steel

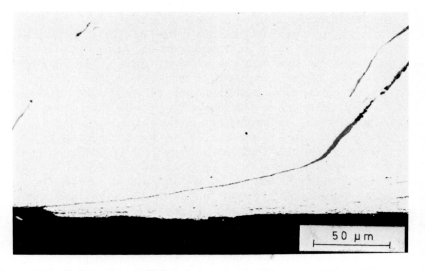

Figure 7.14   Deformation of MnS inclusions in the flow-zone of free-cutting steel chip.   (B.W. Dines[9])

The contact length on the rake face of the tool is shortened by the presence of MnS.[9]   Separation of the chip from the tool, to which it is bonded, requires fracture, and the weak interfaces between the sulphide ribbons and the steel form nuclei for this fracture.   The shorter contact length results in thinner chips and lower tool forces and power consumption, *Figures 7.16*[9] and *7.17*.[17]

After cutting, layers of MnS are often found covering parts of the tool surface.   When using the steel cutting grades of carbide tool, sulphide is found covering the contact area on the tool rake face.   This may prevent almost entirely the formation of a built-up edge at low cutting speed (see machining chart, *Figure 7.18*), and at high speed MnS seems to act as a lubricating layer interposed between tool and work material on the rake face.   The presence of the sulphide layers can be investigated by electron probe analysis, but are readily demonstrated by sulphur prints of the tool rake surface.   *Figures 7.19a* and *b* show a photomicrograph of the contact area on the rake surface of a steel cutting grade of carbide tool after cutting a free-cutting steel, and the sulphur print of this region, which shows a high concentration of sulphide over the whole contact area.   The mechanism by which the MnS particles are deposited on

Figure 7.15   Scanning electron micrograph of MnS in flow-zone of chip after etching deeply in HNO₃.   (B.W. Dines[9]*)*

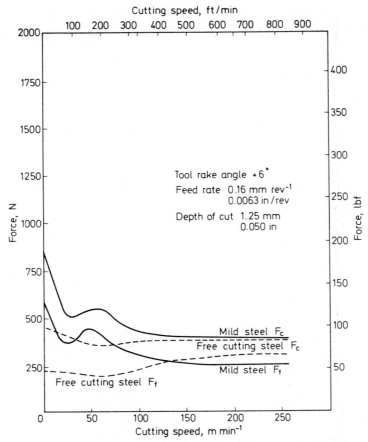

Figure 7.16   Tool forces vs *cutting speed — carbon and free-cutting steels.   (From data in Williams, Smart and Milner*[1]*)*

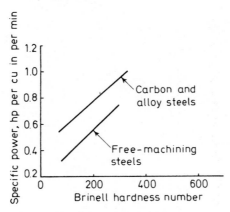

Figure 7.17   Power consumption when cutting carbon and free-cutting steels.[17] *(By permission from Metals Handbook, Volume 1, Copyright American Society for Metals, 1961)*

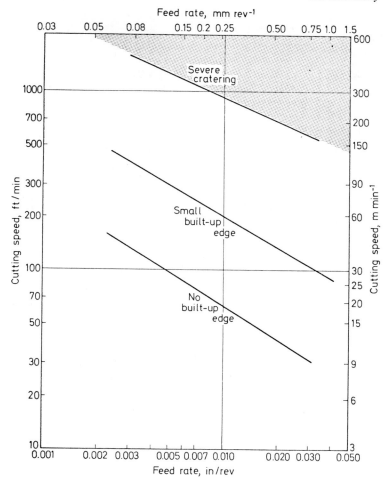

Figure 7.18 'Machining chart' for steel-cutting grade of carbide used for cutting a low carbon free-cutting steel

the tool surface is not certain, but it has been suggested that they are extruded onto the tool surface from their 'sockets' in the steel like tooth paste from a tube.

When using high speed steel tools or cemented carbides of the WC–Co class for machining free-cutting steels, the sulphides are not found on the contact area of the tool face, except at very low cutting speeds, although they often cover areas of the tool beyond the contact region. It is only when using the steel cutting carbide grades that formation of a large built-up edge is prevented. The retention of a sulphide layer at the interface with this class of tool must be due either to chemical bonding of MnS to the cubic carbides, or to mechanical keying to the uneven surfaces worn on the rake face of these tools.[10]

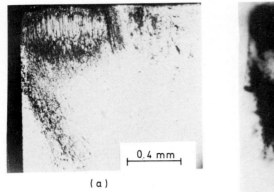
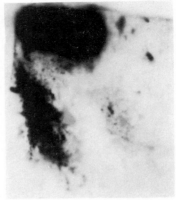

(a)

(b)

Figure 7.19   (a) Sulphides on rake face of steel-cutting grade carbide tool after cutting free-cutting steel at high speed.   (b) Sulphur print of the same tool[10]

The role of lead in the machining process has been studied less than that of sulphur.   It is used as additions of about 0.25% to both free-cutting steels and to steels of normal sulphur content and permits higher rates of metal removal, gives better surface finish and better control of the chips, without serious detriment to the mechanical properties of the steel.   The mechanism by which lead achieves these advantages is uncertain and, under some cutting conditions, it may result in new and more rapid forms of tool wear.   The value of leaded steel as a solution to particular production problems has to be investigated for each case, in the absence of basic knowledge of the part which lead plays in cutting.

Variable machinability of non free-cutting steel

High sulphur content in steel promotes machinability, but very low sulphur content makes machining more difficult.   The rates of flank wear increase with decreasing sulphur content, *Figure 7.20,* as has been demonstrated in a variety of tests,[11]   and steels which have been cleaned of non-metallic inclusions by electro-slag remelting (ESR) are reported to be more difficult to machine.   In some cases small amounts of sulphur have been re-introduced into these steels, e.g. 0.015%S, to reduce costs of machining.

When machining non free-cutting steels the tool wear rates, and the permissible rates of metal removal, are found to vary widely for different consignments of steel conforming to the same standard specification.[11] *Figure 7.21* shows flank wear rate *vs* cutting speed curves for bars of a medium carbon steel from different heats.   The differences may be very large, and pose severe problems in the engineering industry for those involved in planning machining, particularly on transfer lines and highly automated operations.   Not all the causes of the different behaviour

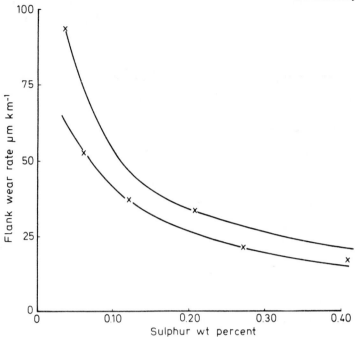

Figure 7.20 Influence of sulphur in steel on the rate of flank wear. (Courtesy Naylor, Llewellyn and Keane,[11] *British Steel Corporation, Swinden Laboratories)*

have been determined, but a major role must be attributed to non-metallic inclusions which behave plastically during machining in a similar way to MnS.[10,12,13] Most oxides in steel are present as inclusions which are undeformed during machining, even in the flow-zone, but certain inclusions, particularly calcium aluminium silicates, are drawn out long and thin. When using the steel cutting grades of carbide tools, they attach themselves, like MnS, to the contact area on the tool rake face to form a glassy barrier between tool and work material. *Figure 7.22* shows such a layer on the rake face and *Figure 7.23* shows a section through a rather thick silicate layer. With steels containing such inclusions, the rates of flank and crater wear, in high speed machining with steel cutting grades of carbide tool, are much lower and the permissible cutting speeds much higher.[13]

This is a good example of how 'specific' machinability can be. Because these inclusions behave in a plastic manner in the extreme conditions of the flow-zone, and become attached to some constituent of the steel-cutting grades of carbide when cutting at high speed, the performance of the tools is greatly increased in operations of economic importance for the engineering industry. Apart from increased rate of metal removal and longer tool life, surface finish may be improved and chip shape altered beneficially. The same work materials, however, may show no

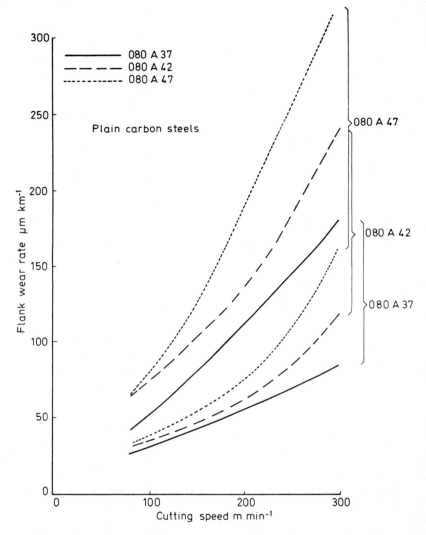

Figure 7.21   Variability of flank wear rate on carbon steels of the same specification. (Courtesy Naylor, Llewellyn and Keane,[11] British Steel Corporation, Swinden Laboratories)

improvement in machinability when cutting with a WC–Co grade of carbide, or with high speed steel tools, nor at low cutting speeds where a built-up edge is formed with any class of tool material. **Machinability is not a property of a material, but a mode of behaviour of the material during cutting, and assessments of machinability should, therefore, specify the general conditions of cutting for which they have validity.**

The oxides and silicates present as inclusions in a steel are mainly the result of the deoxidation practice in the final stages of steel making, which varies considerably in different steel works and for production of

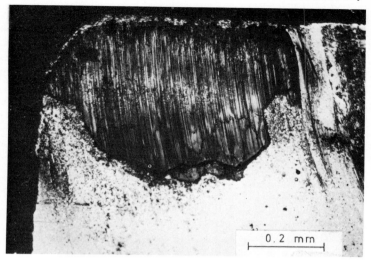

Figure 7.22   Silicate layer on rake face of steel cutting grade of carbide used to cut steel at high speed

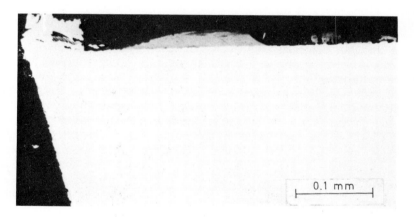

Figure 7.23   Section through silicate layer on rake face of tool[10]

steels for different purposes.[13]   This accounts for much of the variability in machining quality of carbon and low alloy steels.   If widespread use could be made of the deoxidation practice employed in making those steels, at present available, which contain inclusions conferring good machinability, this would be a great advantage to machinists in terms of consistently high rates of metal removal in high speed cutting.   Steels deoxidised with large amounts of aluminium are known to give short tool life.   This may be due to the presence of abrasive Al_2O_3 particles in the steel, but it seems more likely that the cause is the absence of beneficial plastic inclusions in steels de-oxidised in this way.   Deliberate attempts have been made to devise a de-oxidation practice which produces inclusions of an optimum character for machining. It is suggested that they are effective with an optimum degree of plasticity and cease to function at their melting point. The 'specially deoxidised' steels usually contain calcium aluminium silicates and Opitz and König[14] give results which show a typical composition of 55% CaO, 30% Al_2O_3, 15% SiO_2 and a melting point of 1380 °C. However, good machinability is rarely the major quality required of a steel, and deoxidation practice has to be designed to ensure correct response to heat treatment and the optimum properties in the final product, and these qualities may demand other types of inclusion. Whether or not the 'special deoxidation' practice can be generally adopted, it is certain that there is scope for adjustment of steel making practice to produce steels capable of more consistent performance during machining at high rates of metal removal, without resorting to the addition of large percentages of sulphur.

Austenitic stainless steel

Austenitic stainless steels are generally regarded as more difficult to machine than carbon or low alloy steels. They bond very strongly to the tool during cutting, and this bonding is more obvious than when cutting other steels because the chips more often remain stuck to the tool after cutting. When the chip is broken away it may bring with it a fragment of the tool, particularly when cutting with cemented carbides, giving poor and erratic tool performance.

The tool forces are not greatly different from those when cutting normalised medium carbon steel (*Figure 7.24*). The temperature pattern imposed on the tool is of the same general character as when cutting other steels, with a cool region at the cutting edge. The cutting speed to achieve any temperature is rather lower than with a medium carbon steel. A built-up edge is formed in a cutting speed range somewhat lower than with carbon steels, and has rather a different character, being more like an enlarged flow-zone (*Figure 7.25*). Cratering by the hot shearing mechanism occurs when cutting with high speed steel tools at speeds lower than those for medium carbon steels (*Figures 6.9* and *6.10*). The relatively high temperatures generated restrict the rates of metal removal and, at a feed rate of 0.25 mm (0.01 in) per rev, cutting speeds with high speed steel tools are generally lower that 25–30 m min^{-1} (80–100 ft/min). Flank wear when cutting austenitic stainless steels is characteristically very smooth and even, with high speed steel or with carbide tools, and increases regularly as cutting speed is raised. The range of cutting speed and feed for a carbide tool are shown in the machining chart, *Figure 7.26*.

Austenitic stainless steels are strongly work-hardening, and particular problems arise when cutting into a severely work-hardened surface, such as that left by a previous machining operation with a badly worn tool. The use of sharp tools and a reasonably high feed rate are two recommendations for prevention of damage to tools caused by this work-hardening.

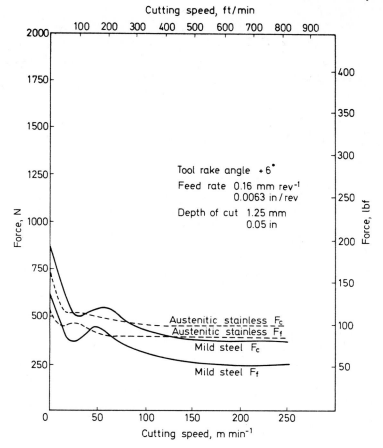

Figure 7.24 Tool forces vs *cutting speed – austenitic stainless steel*

Free-cutting austenitic stainless steels are available to increase tool life and metal removal rates and to facilitate difficult machining operations. These have high sulphur contents, and, as with the free-cutting ferritic steels, their improved machinability is associated with the plastic behaviour of the sulphides in the flow-zone. The use of free-cutting stainless steels is restricted, because the introduction of large numbers of sulphide inclusions reduces corrosion resistance under some conditions. An alternative free-cutting additive is selenium, which forms selenide particles which behave like sulphides during machining. Selenides are advocated because they are said to be less harmful to corrosion resistance.

Cast iron

Flake graphite cast irons are considered to have very good machining qualities. A major reason for the continued large-scale use of cast iron in

Figure 7.25   Section through built-up edge on tool used to cut austenitic stainless steel.   (P.K. Wright)

engineering is not only the low cost of the material and the casting process, but also the good economics of machining the finished component. By nearly all the criteria it has good machinability – low rates of tool wear, high rates of metal removal, relatively low tool forces and power consumption. The surface of the machined cast iron is rather matt in character, but ideal for many sliding interfaces. The chips are produced as very small fragments which can readily be cleared from the cutting area even when machining at very high speeds. It is a somewhat dirty and dusty operation, throwing a fine spray of graphite into the air, so that some protection for the operators may be required.

As when cutting other materials, there is a great difference between the behaviour of cast iron when sheared on the shear plane and at the tool work interface. The most important characteristic is that fracture on the shear plane occurs at very frequent intervals, initiated by the graphite flakes, so that the chip is composed of very small fragments a few millimetres in length. Because the chips are not continuous, the length of contact on the rake face is very short, the chips are thin and the cutting force and power consumption are low. The cutting force is low also because graphite flakes are very weak and may be relatively large so that one flake may extend an appreciable way across the shear plane. *Table 7.2* shows values for the cutting force (F_c) for a typical pearlitic iron in comparison with steels under one set of cutting conditions. This aspect of machinability is influenced by the grade and composition of the cast iron. Low strength irons, the structure of which consists mainly of ferrite and graphite, are the most machinable, permitting the highest rates of metal removal. Permissible speeds and feeds are somewhat lower for pearlitic irons, and decrease as the strength and hardness are raised. The

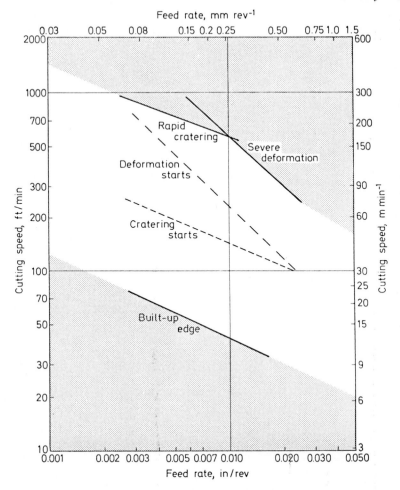

Figure 7.26 'Machining chart' for steel cutting grade of carbide ('Wimet XL3') used for cutting austenitic stainless steel

highly alloyed irons and chilled irons, with very little graphite and containing large amounts of iron carbide (Fe_3C) and other metal carbides, become very difficult to machine. Chilled iron rolls may be machined only at speeds of the order of 3–10 m min^{-1} (10–30 ft/min) even with cemented carbide tools.

The majority of engineering cast irons are of the ferritic or pearlitic types, and their behaviour on the shear plane during cutting can be predicted quite well from their strength and lack of ductility as measured in standard laboratory tests. Their behaviour at the tool/work interface is, however, less 'conventional'. Graphite might be expected to act as a lubricant and to inhibit seizure at the tool/work interface, but there is no evidence that it acts in this way. When cutting with carbide or high

Table 7.2   Cutting forces — pearlitic, flake-graphite cast iron

Cutting speed		Cast iron				Mild steel			
		F_c		F_f		F_c		F_f	
m min^{-1}	ft/min	N	lbf	N	lbf	N	lbf	N	lbf
30	100	222	50	232	52	520	115	356	80
61	200	245	55	285	64	490	110	364	82
91	300	245	55	320	72	445	100	325	75
122	400	267	60	338	76	422	95	313	70

NOTE.   Feed rate: 0.16 mm rev^{-1} (0.0063 in/rev)
Depth of cut: 1.25 mm (0.05 in)

speed steel tools a built-up edge is formed which persists to higher cutting speeds than with steels.   *Figure 6.38* is a machining chart for a pearlitic iron cut with a WC–Co tool, showing the region in which the built-up
edge persists.   The built-up edge changes shape as speeds and feeds are
raised, and eventually disappears when a form of cratering appears on the
tool.   *Figure 7.27* shows a section through a built-up edge seized to the
rake face and to the worn flank.   This consists of fine fragments of the
metallic parts of the cast iron structure, extremely severely plastically
strained and welded together, the cementite and other constituents usually
being so highly dispersed that they cannot be resolved with an optical
microscope.   No graphite is identified in the built-up edge – the top part
of *Figure 3.8* shows the structure observed in a polished but unetched
section.   Graphite is probably present, broken up into very thin, fragmented layers, since a black deposit is formed when the built-up edge is
dissolved in acid.

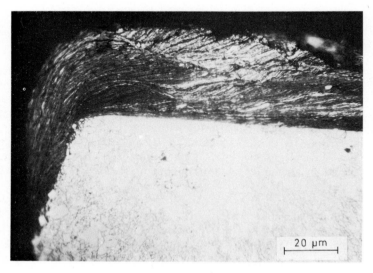

*Figure 7.27   Section through built-up edge on carbide tool used to cut pearlitic
flake-graphite cast iron*

Thus, under the compressive stress and conditions of strain at the tool surface, flake graphite ferritic and pearlitic cast irons behave as plastic materials. The feed force (F_f) is thus often higher than the cutting force (F_c) as shown in *Table 7.3,* and nearer to the value of F_c when cutting steel. It is normal practice when machining with either high speed steel or cemented carbide tools, to cut cast iron under conditions where a built-up edge is formed, and very good tool life can be achieved. With a discontinuous chip, the built-up edge is more stable and is less frequently detached from the tool even when an interrupted cut is involved. The wear is by attrition (*Figures 3.8* and *3.9*) and the longest tool life is achived with tungsten carbide-cobalt tools of fine grain size (*Figure 6.42*). At higher rates of metal removal the built-up edge disappears, and, to resist cratering and diffusion wear on the flank, a fine grained steel-cutting grade of carbide tool can be used, containing small amounts of TiC and TaC, or TiC coated tools may also give low wear rates.

Investigations so far suggest that the temperature distribution in tools used to cut cast iron is different from that when cutting steel. In the absence of a continuous chip, the highest temperatures are observed in the region of the cutting edge. With high compressive stress and temperature both in this region, the upper limit to the rate of metal removal occurs when the tools are deformed at the edge, see machining chart, *Figure 7.26.*

Ceramic tools (alumina), having excellent resistance to diffusion wear, and greater high temperature strength than cemented carbide, can be used to cut flake graphite pearlitic iron at over 700 m min^{-1} (2 000 ft/min) compared with a maximum of about 180 m min^{-1} (500 ft/min) with cemented carbide tools.

Spheroidal graphite (SG) irons have better mechanical properties than flake graphite irons, and in recent years have replaced them in many applications. In the SG irons the graphite is present as small spheres instead of flakes, but during cutting they behave in a very similar way to the flake graphite irons, and can generally be machined using very similar techniques. The graphite spheres act to weaken the material in the shear plane and initiate fracture, but are rather less effective in this respect than flake graphite. The chips are formed in rather longer segments, but these are weak, easily broken and much nearer in character to flake graphite iron, than to steel chips. One problem which is sometimes encountered is that, with ferritic SG iron, the flow-zone material is extremely ductile and may cling to the clearance face of the tool when cutting at high speeds, causing very high tool forces, high temperatures and poor surface finish. This problem can be largely overcome by using high clearance angles on the tools.

Nickel and nickel alloys

Although nickel has a lower melting point (1 452 °C) than iron (1 535 °C), the metal and its alloys are, in general, more difficult to machine than iron and steel. Nickel is a very ductile metal with a face-centred cubic structure and, unlike iron, it does not undergo transformations in its basic crystal structure up to its melting point. Commercially pure nickel has poor machinability on the basis of almost all the criteria. Tool life tends to be short and the maximum permissible rate of metal removal is low, the tools failing by rapid flank wear and deformation of the cutting edge at relatively low cutting speeds. A recommended cutting speed is 50 m min^{-1} (150 ft/min) with high speed steel tools when turning at a feed rate of 0.4 mm (0.015 in) per rev. Tool forces are higher than when cutting commercially pure iron (*Figure 7.28*), the contact area on the rake face being very large, with a small shear plane angle and very thick chips. As with iron and other pure metals, no built-up edge is formed, and the tool forces decrease steadily as the cutting speed is

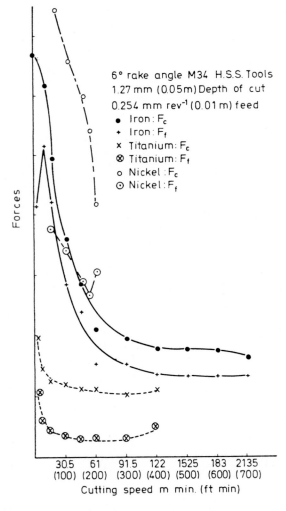

6° rake angle M34 H.S.S. Tools
1.27 mm (0.05m) Depth of cut
0.254 mm rev^{-1} (0.01 m) feed
● Iron: F_c
+ Iron: F_f
× Titanium: F_c
⊗ Titanium: F_f
o Nickel: F_c
⊙ Nickel: F_f

Forces

30.5   61   91.5   122   152.5   183   213.5
(100) (200) (300) (400) (500) (600) (700)
Cutting speed m min. (ft min)

Figure 7.28   Tool forces vs *cutting speed − comparison of iron, nickel and titanium*[15]

(a)

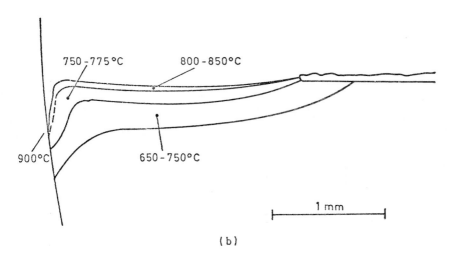

(b)

Figure 7.29   (a) Section through high speed steel tools used to cut commercially pure nickel, etched to show temperature distribution (b) Temperature contours derived from (a)[15]

raised, the contact area becoming smaller and the chip thinner, but over the whole speed range the forces are relatively high.

The most important aspect of the behaviour of nickel during cutting, which leads to high rates of tool wear and low rate of metal removal, appears to be the high temperatures generated in the flow-zone, and a characteristic adverse distribution of temperature in the tools, which is very different from that when cutting iron and steel.[15]   *Figure 7.29a* shows an etched section through a high speed steel tool used to cut commercially pure nickel at 45 m min⁻¹ (150 ft/min) at a feed rate of 0.25 mm (0.010 in) per rev feed. During manual disengagement of the tool, separation took place along the shear plane, leaving the chip very strongly bonded to the tool. The flow-zone, which is the heat source, is very clearly delineated adjacent to the tool rake face. The derived temperature gradient from this tool is shown in *Figure 7.29b*. *Figures 7.29a* and *b* should be compared with *Figures 5.3a* and *b* to contrast the characteristic temperature distributions in tools used to cut nickel and iron. There are two major differences, (1) temperatures over 650 °C appear at much lower speeds when cutting nickel, and (2) the cool region at the tool edge is not present when cutting nickel. The difference in temperature distribution is seen also if the rake surfaces of tools used to cut nickel and iron are compared (*Figures 7.30* and *5.5*). Temperature is seen to be high along the main cutting edge when cutting nickel, but not on the end clearance face, a location where very high temperatures led to deformation of the tool used for cutting iron. Consequently, tools used for cutting nickel tend to be deformed along the main cutting edge, where both compressive stress and temperature are high even at relatively low cutting speeds. Once the tool edge has deformed and a wear land has been started, a new heat source develops at the flank wear land and may result in rapid collapse of the tool.

Cemented carbide tools, with their higher compressive strength at high temperature, can be used for cutting nickel and its alloys at much higher

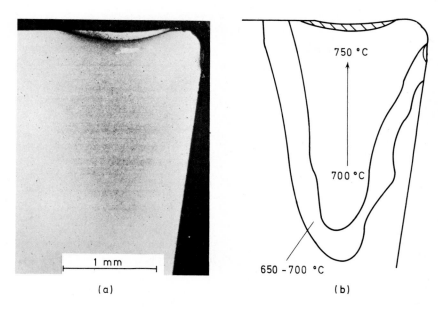

(a)   (b)

Figure 7.30 *(a) Rake face of tool used to cut nickel, etched to show temperature distribution. (b) Temperature contour derived from (a)*[15]

speeds than high speed steel tools. Carbide tools wear mainly on the flank by a diffusion or deformation mechanism, cratering not being a major problem. Cemented carbide tools are not, however, generally recommended for cutting commercially pure nickel, since the very strong bonding of the nickel chips to the tool surface often leads to damage to the tool when the chips are removed.

The addition of alloying elements to nickel affects its machining qualities in ways similar to those discussed for iron and steel. Even when the alloying additions result in considerable strengthening of the nickel, the cutting forces are often reduced because the contact length on the rake face is smaller, the shear plane angle is larger and the chip thinner. Rather meagre evidence so far available suggests that nickel alloys in general impose on the tools a temperature distribution similar to that observed when cutting nickel, so that tool life and the maximum rate of metal removal are controlled by the same mechanisms. The effect of alloying elements is to reduce the cutting speed at which high temperatures are generated and to increase the localised stresses on the tool, thus increasing tool wear rates and reducing maximum permissible rates of metal removal.

The creep-resistant nickel-based alloys, used in the aero-space industry, are some of the most difficult materials to machine. These alloys are strengthened by a finely dispersed second phase, as well as by solid solution hardening. A built-up edge is formed when cutting these two-phased alloys at low cutting speeds (*Figure 7.31*). As the speed is raised the

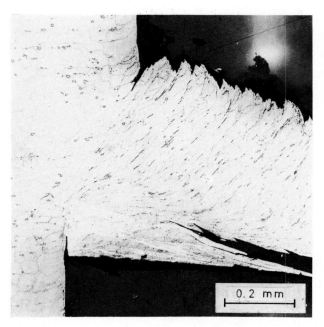

Figure 7.31   Built-up edge when cutting creep-resistant two-phased nickel alloy, quick-stop

174

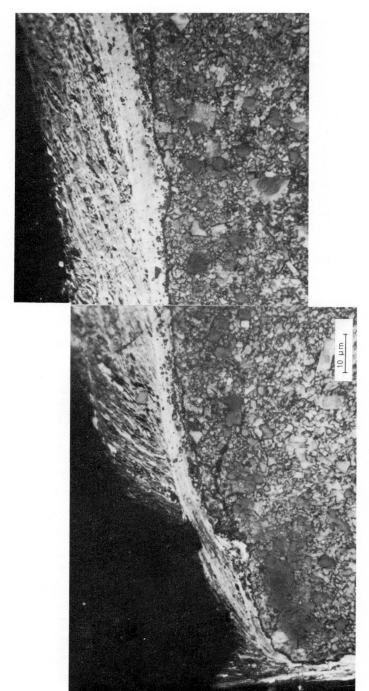

Figure 7.32 Section through cutting edge of carbide tool used to cut cast creep resistant nickel alloy at 17 m min⁻¹ (50 ft/min) cutting speed[3]

built-up edge disappears but very high temperatures are generated even at relatively low speeds in the flow-zone at the tool/work interface. The temperatures are often high enough to take into solution the dispersed second phase in the nickel alloy, and may be well over 1 000 °C. Because these creep-resistant alloys are metallurgically designed to retain high strength at elevated temperatures, the stresses in the flow-zone are very high. The result is a destruction of the cutting edge under the action of shear and compressive stresses acting at high temperature. *Figure 6.11* is a section through the cutting edge of a high speed steel tool used to cut one such wrought alloy at 10 m min⁻¹ (30 ft/min) and shows the tool material being sheared away. For many operations on nickel-based alloys, cemented carbides are the most efficient tool material in current use. Fine grained WC–Co alloys are most commonly used, the steel cutting grades of car-bides rarely showing any advantage. When cutting the most advanced of the 'aero-space alloys', however, the inadequacy of cemented carbide tools becomes apparent. The tearing apart of a carbide tool used to cut one of the most creep-resistant cast nickel-based alloys is shown in *Figure 7.32,* the cutting speed in this case being only 16 m min⁻¹ (50 ft/min).

Thus, with the most highly creep resistant alloys, on which much mach-ining has to be done for aero-engine components, it may become almost impossible to cut at economic speeds, even with carbide tools, because of the severity of stress and temperature in the flow-zone where tool and work are seized. Much effort has been devoted to the development of new methods and tools capable of higher rates of metal removal with these alloys. Carbide tips with sintered layers of 'Borazon' bonded to the surface have been used, and it is claimed that, with these, cutting speeds many times higher than with carbide tools can be achieved with longer tool life, e.g., 180 m min⁻¹ (600 ft/min) with 'Borazon' compared with 30 m min⁻¹ (100 ft/min) with cemented carbide tools. In spite of their very high cost (more than 60 times that of carbide tools) such a great difference in performance could make them economic to use.

A method of hot-machining has been proposed and used for certain operations, the material to be machined being rapidly heated as it approached the cutting tool, usually by a plasma torch. Under the right conditions, this makes possible the use of ceramic tools for cutting these creep-resistant alloys at several times the rate possible with carbide tools. The explanation for the improved performance is not certain, but, with-out the pre-heating of the work material, ceramic tools are not normally successful in cutting nickel-based alloys, the tools failing suddenly after a short life. The range of shapes and applications where this method can be applied is limited.

Titanium and titanium alloys

Titanium and its alloys are generally regarded as having rather poor mach-inability. The melting point of Ti is 1 668°C, and it is a ductile metal with a close-packed hexagonal structure at room temperature, changing to body-centred cubic at 882ᵛC. The commercially pure metal is available in a range of grades depending on the proportion of the elements carbon, nitrogen and oxygen, the hardness and strength increasing and the ductility decreasing as the content of these elements is raised.

The machining characteristics of titanium are different in several respects from those of the other pure metals so far considered, and by several of the criteria, it cannot be said to have poor machinability. Tool life is terminated by flank wear and/or deformation of the tool, and the rates of metal removal for a reasonable tool life are lower than when cutting iron. The tool forces and power consumption, however, are much lower than when cutting iron, nickel or even copper, especially in the low-speed range, as shown in *Figure 7.28*. These low tool forces are associated with a much smaller contact area on the rake face of the tool than when cutting any of the other metals discussed except magnesium. Because of the small contact area, the shear plane angle is large and the chips are thin, often not much thicker than the feed. The chips are continuous but are typically segmented, and, with titanium alloys, the segmentation becomes very marked, narrow bands of intensely sheared metal being separated by broader zones only lightly sheared (*Figure 7.33*). No built-up edge is formed when cutting commercially pure titanium, and the flow-zone at the rake face tends to be very thin – usually less than 0.012 mm (0.0005 in) and often much thinner than this.

The main problems of machining titanium are that the tool life is short and permissible rates of metal removal are low, in spite of the low tool forces. It is the high temperatures and unfavourable temperature distribution in tools used to cut titanium which are responsible for this.[15] The temperatures in the flow-zone are higher than when cutting iron at the

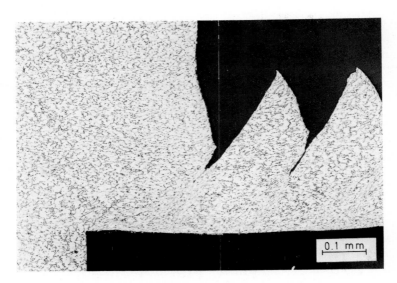

Figure 7.33   Section through forming titanium alloy chip, quick-stop.   (R.M. Freeman[16])

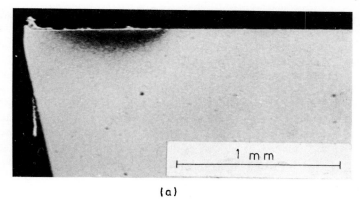

(a)

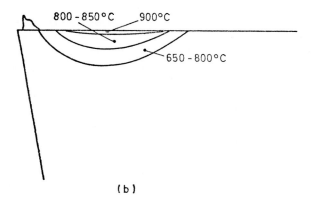

(b)

Figure 7.34 (a) Section through high speed steel tools used to cut commercially pure titanium, etched to show temperature distribution. (b) Temperature contours derived from (a)

same speed, for example the maximum temperature on the rake face of a tool was 900 °C after cutting a commercially pure titanium at 91 m min^{-1} (300 ft/min) and 650 °C after cutting iron at this speed under the standard cutting conditions. Temperature gradients in tools used to cut titanium are shown in *Figure 7.34* and should be compared with those for cutting iron (*Figure 5.3*) and nickel (*Figure 7.29*). The temperature distribution is more like that when cutting iron, but the cool zone close to the edge is very narrow, and the high temperature region is much closer to the tool edge. The total contact length is very short and the heated region does not extend far along the rake face. Thus, although the tool forces are low, the stress on the rake face is high, and a highly stressed region near the tool edge is at a high temperature. This leads to deformation of the tool edge and rapid failure, with the formation of a new heat source on the deformed and worn flank. Frequently failure is initiated at the nose radius of the tool.

The temperature gradients in tools used to cut titanium alloys are similar in character to those found when cutting the commercially pure metal,[16] and different from those when cutting steels or nickel-based alloys. In general, the effect of alloying additions is to raise the temperature for any set of cutting conditions, and therefore, to reduce the permissible cutting speed. When cutting commercially pure titanium, the influence of increasing amounts of the interstitial impurity elements, carbon, nitrogen and oxygen, is very pronounced. In one series of experiments,[16] an increase in oxygen content from 0.13% to 0.20% reduced the cutting speed required to produce a temperature of 900 °C, in the tool from 91 m min[-1] (300 ft/min) to 53 m min[-1] (175 ft/min).

With alloys containing a second phase, the temperature increase for any cutting speed is much more marked.[16] Using the standard test conditions, tools used to cut an alloy containing 6% Al and 4% V were heated on the rake face to over 900 °C at a cutting speed of 19 m min[-1] (60 ft/min). When cutting a commercial alloy with 11% Sn, 2.25% Al and 4% Mo, high speed steel tools failed due to stress and temperature after cutting for only 30 s at a speed of 12 m min[-1] (40 ft/min). In failure of high speed steel tools not only is the edge deformed downward under compressive stress, but the heated high speed steel is sheared away to form a crater on the rake face, as was observed when cutting steel.

Apart from deformation, diffusion wear seems to be the main process responsible for the wear both of high speed steel and of carbide tools when cutting titanium alloys. With cemented carbide tools, longer life is achieved with the use of the WC–Co alloys than with the steel-cutting grades containing TiC and TaC. The introduction of TiC, which is so strikingly successful in combating diffusion wear when cutting steel, has an adverse effect in relation to diffusion wear when machining titanium and its alloys. There is evidence that the cubic carbide grains containing TiC are lost more rapidly by diffusion into titanium flowing over the tool surface than are the WC grains.[16] Resistance to diffusion wear and resistance to deformation at high temperatures make the WC–Co grades of carbide useful for cutting titanium alloys. Even with these, the cutting speeds which can be used for machining the more creep-resistant alloys are low, e.g., 30 m min[-1] (100 ft/min).

Zirconium

It is of interest that the machining behaviour of commercially pure zirconium is very similar indeed to that of titanium. The contact area on the rake face of the tool is short, the shear plane angle high and the chips are thin. The same sort of temperature pattern is imposed on the tool as when cutting titanium, and these are the only two high melting-point metals investigated so far for which the temperature gradients in the cutting tools are similar in character. The very close similarity of these two metals in structure and properties is paralleled by their machining qualities. There is some hazard in machining zirconium because fine swarf may ignite and precautions specified by the material suppliers should be observed.

General observations on machinability

In this chapter the machining qualities of some of the more commonly used metals and alloys have been described and discussed. It is clear that machining behaviour is complex, not easily described within a short compass and it cannot meaningfully be evaluated by a single measurement. Useful *ad hoc* tests can be specified for prediction of tool life, rates of metal removal or power consumption under particular sets of operating conditions, but these cannot be regarded as evaluations of machinability, valid for the whole range of operations encountered industrially. The results of such tests should always be accompanied by a statement of the machining operation used and the critical test conditions. Progress can be made towards a more basic understanding of the machining qualities of metals and alloys by studying their behaviour during machining, investigating both the changes that take place in the material as it passes through the primary shear plane, and the very different changes in the secondary shear zone at the tool-work interface. The former can, to some extent, be directly related to properties measured by standard laboratory mechanical tests, but behaviour in the secondary shear zone can be investigated only by observations of the machining process, since it cannot readily be simulated by model tests.

It is with the higher melting-point metals and alloys that difficult problems most frequently arise in industrial machining practice, and, for these materials, the temperature and temperature distribution in the secondary shear zone (the flow-zone) play an important role in almost every aspect of machinability. The evidence of laboratory experiments shows that each of the major metals imposes on the tools a characteristic temperature pattern which differs greatly for different metals. The reason why a particular temperature pattern is associated with a particular metal is an interesting subject for research. Fortunately for industrial practice, the temperature pattern associated with iron and steel is favourable for high rates of metal removal. The temperature distribution is probably associated with the characteristic pattern of flow adopted by the metal as it flows around the cutting edge and over the tool surfaces, since the temperature is dependent on the energy expended in deforming the metal in the flow-zone, and the quantity of material flowing through this zone. Much study of flow patterns will be required before a satisfactory explanation can be given. This is a most difficult region to study because of the small size and inaccessibility of the critical volume of metal, but a better understanding of behaviour in the flow-zone is an essential pre-requisite for comprehension of machinability. These investigations have also theoretical interest in relation to the behaviour of materials subjected to extreme conditions of strain and strain rate.

References

1. WILLIAMS, J.E., SMART, E.F. and MILNER, D.R., *Metallurgia*, **81**, 3, 51, 89 (1970)
2. SULLY, W.J., *I.S.I. Special Report*, **94**, 127 (1967)
3. TRENT, E.M., *I.S.I. Publication*, **126**, 15 (1970)
4. DAVIES, D.W.. *Metals Technology*, **3**, Parts 5 and 6, 272 (1976)
5. TRENT, E.M., *Proc. Int. Conf. M.T.D.R. Manchester 1967*, 629 (1968)
6. MARSTON, G.J. and MURRAY, J.D., *J.I.S.I.*, 568 (1970)
7. SHAW, M.C., SMITH, D.A. and COOK, N.H., *Trans. Am. Soc. Mech. Eng.*, **83B**, 181 (1961)
8. MOORE, C., *Proc. International Conference MTDR*, 929 (1968)
9. DINES, B.W., *Ph.D. Thesis*, University of Birmingham (1975)
10. TRENT, E.M., *I.S.I. Special Report*, **94**, 77 (1967)
11. NAYLOR, D.J., LLEWELLYN. D.T. and KEANE, D.M., *Metals Technology*, **3**, Parts 5 and 6, 254 (1976)
12. PIETIKÄINEN, J., *Acta Polytechnica Scandinavica*, No.91, (1970)
13. OPITZ, H., GAPPISCH, M. and KÖNIG, W., *Arch. für das Eisenhütten-wesen*, **33**, 841 (1962)
14. OPITZ, H. and KÖNIG, W., *I.S.I. Special Report*, **94**, 35 (1967)
15. SMART, E.F. and TRENT, E.M., *Int. J. Prod. Res.*, **13**, 3, 265 (1975)
16. FREEMAN, R., *Ph.D. Thesis*, University of Birmingham (1975)
17. *A.S.M. Handbook*, 8th edition, Vol.1, 306 (1961)

8

COOLANTS AND LUBRICANTS

A tour of most machine shops will demonstrate that some cutting operations are carried out dry, but in many other cases, a flood of liquid is directed over the tool, to act as a coolant and/or a lubricant. These cutting fluids perform a very important role and many operations cannot be efficiently carried out without the correct type of fluid.[1]   They are used for a number of objectives (1) to prevent the tool, workpiece and machine from overheating; (2) to increase tool life; (3) to improve surface finish, and (4) to help clear the swarf from the cutting area.   Many machine tools are fitted with a system for handling the cutting fluids — circulating pumps, piping and jets for directing the fluids to the tool, and filters for clearing the used fluid.

A very large number of cutting fluids is available commercially from which the machinist selects the one most suitable for a particular application.   With very little guidance from theory, both the development of cutting fluids and their selection depend on a vast amount of empirical testing.   A successful fluid must not only improve the cutting process in one of the ways specified, but must also satisfy a number of other requirements.   It must not be toxic or offensive to the operator; it must not be harmful to the lubricating system of the machine tool; it should not promote corrosion or discolouration of the work material, and should preferably afford some corrosion protection to the freshly cut metal surface; it should not be a fire hazard, and it should be as cheap as possible.

There are two major groups of cutting fluid — the water-based fluids and the neat cutting oils.   The former act mainly as coolants and consist of an emulsion of mineral oil in water, usually in proportions between 1 to 10 and 1 to 30 of oil to water.   The latter are basically mineral oils and their action is mainly as lubricants.   There are many variants of both types.   Both the emulsions (which are commonly known as 'soluble oils') and the neat oils may contain chlorine and sulphur additives which effect improvements in lubrication under extremely difficult conditions, and fatty acids are often incorporated in the neat oils. It is not proposed to discuss here the details of the application of each of these types of cutting fluid, but to consider the action of cooling and of lubrication in the light of experimental evidence from laboratory tests, and of knowledge of the conditions existing at the tool-work interface presented in the previous chapters.

Coolants

It is in connection with the machining of steel and other high melting-point metals that the use of coolants becomes essential. Their use is most important when cutting with steel tools, but they are often employed also with carbide tooling. An example of a situation where coolants must be used is on automatic lathes where several tools are used simultaneously or in quick succession to fabricate relatively small components.

In Chapter 5 the two main sources of heat in a cutting operation are discussed — on the primary shear plane and at the tool-work interface (especially in the flow-zone on the tool rake face). The work done in shearing the work material in these two regions is converted into heat, while the work done by sliding friction makes a minor contribution to the heating under most cutting conditions. Coolants cannot prevent the heat being generated, and do not have direct access to the zones which are the heat sources. Heat generated in the primary shear zone is mostly carried away in the chip and a minor proportion is conducted into the workpiece. Water-based coolants act efficiently to reduce the temperature both of the work-piece and of the chip after it has left the tool. The cooling of the chip is of minor importance, but maintaining low temperature in the work-piece may be essential for dimensional accuracy.

The removal of heat generated in the primary shear zone can have little effect on the life or performance of the cutting tools. As has been demonstrated, the heat generated at and near the tool/work interface is of much greater significance, particularly under high cutting speed conditions where the heat source is a thin flow-zone seized to the tool. The coolant cannot act directly on the thin zone which is the heat source, but only by removing heat from those surfaces of the chip, the workpiece, and the tool, which are accessible to the coolant and as near as possible to the heat source. Removal of heat by conduction through the chip and through the body of the workpiece is likely to have relatively little effect on the temperature at the tool/work interface, since both chip and workpiece are constantly moving away from the contact area allowing very little time for heat to be conducted from the source. For example, when cutting at 30 m min^{-1} (100 ft/min) the time required for the chip to pass over the region of contact with the tool is of the order of 0.005 s.

The tool is the only stationary part of the system. It is the tool which is damaged by the high temperatures and, therefore, in most cases, cooling is most effective through the tool. The tool is cooled most efficiently by directing the coolant towards those accessible surfaces of the tool which are at the highest temperatures, since these are surfaces from which heat is most rapidly removed, and the parts of the tool most likely to suffer damage. Knowledge of temperature distribution in the tool can, therefore, be of assistance in a rational approach to coolant application. This is illustrated by experimental evidence from laboratory cutting tests on tools used to cut a very low carbon steel and commercially pure nickel.[2]

Figure 8.1 shows sections through the cutting edge of high speed steel tools used to cut the low carbon steel at high speed *a* dry, *b* flooded

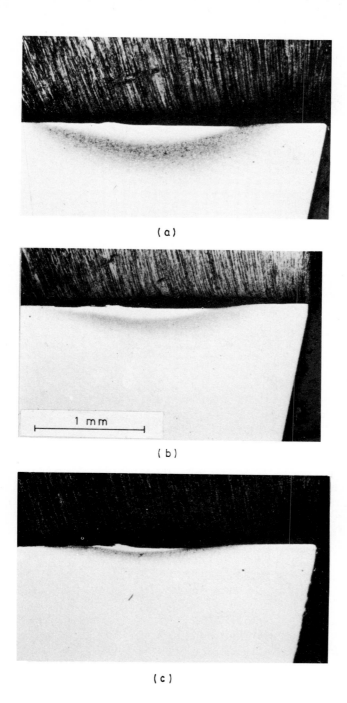

(a)

(b)

(c)

Figure 8.1   (a) Section through high speed steel tool after cutting iron in air at 183 m min⁻¹ (600 ft/min), etched to show temperature distribution.   (b) As (a) for tool flooded with coolant over rake face.   (c) As (a) with jet of coolant directed at end clearance face[2]

184

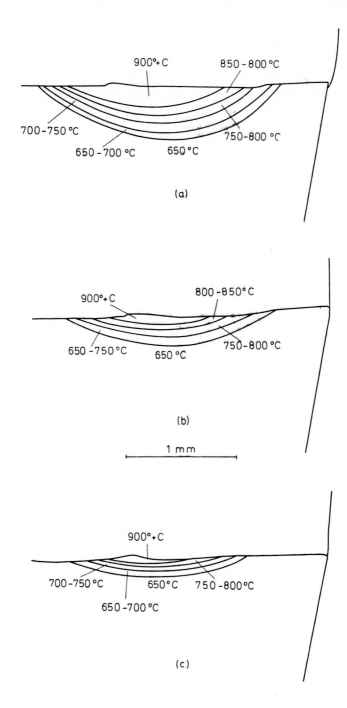

Figure 8.2 (a) Temperature contours derived from Figure 8.1a. (b) Temperature contours derived from Figure 8.1b. (c) Temperature contours derived from Figure 8.1c[2]

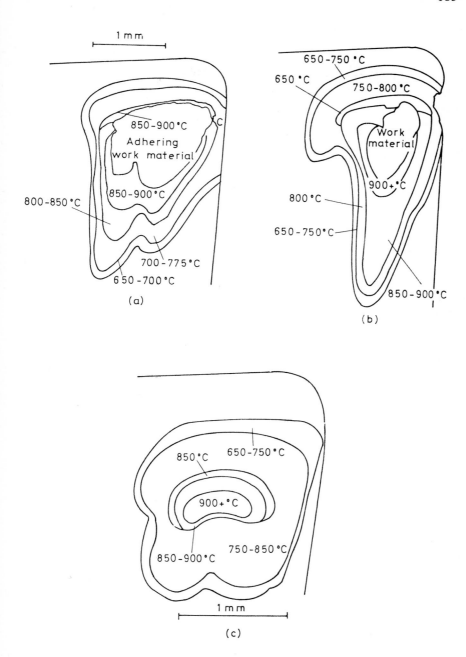

Figure 8.3 (a) Temperature contours on rake face of tool used to cut iron in air, conditions as Figure 8.1a. *(b) Temperature contours on rake face of tool as* Figure 8.1b. *(c) Temperature contours on rake face by tool as* Figure 8.1c[2]

over the chip and tool rake face by a water-based oil-emulsion, and *c*
with a jet of the coolant directed towards the end clearance face of the
tool.   The tools had been sectioned and etched to show the temperature
gradients in the tool by the method described in Chapter 5.   *Figure 8.2*
shows the temperature contours derived from the structures in *Figure 8.1.*
Figure 8.3 shows the temperature contours on the rake faces of tools
used for cutting under the same conditions as those in *Figure 8.2*.   These
illustrate a number of important features relevant to the action of coolants.

First, the coolant application was unable to prevent high temperatures at
the tool/work interface, since heat continues to be generated in the flow-
zone which is inaccessible to direct action by the coolant.   Temperatures
over 900 °C were generated at the hottest part of the rake face of the
tool whether cutting dry, flooded with coolant, or with a jet directed at
the end clearance face.

Second, the action of the coolant reduced the volume of the tool
material which was seriously affected by overheating.   A jet directed to
the end clearance face was much more effective in this respect than
flooding over the rake face.   The temperature gradients within the tool
were much steeper when coolant was used.

Third, the damage to the end clearance face caused by deformation of
the tool when cutting dry or when flooded by coolant from on top, was
prevented when the temperature of this face was reduced by a coolant
jet.   The wider cool zone at the cutting edge in this tool suggests that
the rate of flank wear by diffusion would also be reduced by this
method of cooling.

The cool zone at the cutting edge, which is a feature of tools used to
cut steel, is absent when cutting nickel, as demonstrated in Chapter 7.
The high temperature at the main cutting edge leads to wear at this
edge and failure by deformation.   When cutting nickel, therefore, the
coolant was found to be very effective when directed as a jet on to
the clearance face below the main cutting edge. *Figure 8.4a* shows a
section through a tool used to cut commercially pure nickel dry, while
Figure 8.4b shows the corresponding tool after cutting with a jet of
coolant on the clearance face. The corresponding temperature gradients are
shown in *Figure 8.5*.   These reveal the considerable reduction in temperature in, and wear of, the tool near the edge, achieved by a coolant
directed to the correct part of the tool.

In commercial application, the use of floods of coolant over the rake
face of the tool is simple to arrange, and most machine tools are equipped
with coolant systems designed for this method of application.   When more
efficient cooling is required it is most usual to increase the flow of coolant.   This is not necessarily efficient in relation to improving tool life,
because the coolant so supplied does not necessarily penetrate to the clearance faces of the tool from which heat is most effectively extracted.   A
much smaller amount of coolant directed accurately at the hot spots
might be more effective, although more expensive to engineer.   A 'mist
coolant' spray so directed has been used commercially to solve difficult
problems, as in the cutting of nickel-based alloys.   Another method

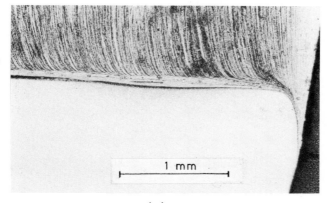

(a)

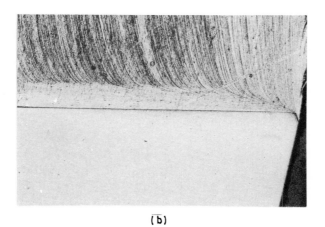

(b)

Figure 8.4   (a) Section through high speed steel tool used to cut nickel in air at 46 m min[-1] (150 ft/min), etched to show temperature distribution.   (b) as (a) with jet of coolant on side clearance face[2]

which is also very effective and clean is the use of CO_2 as a coolant. CO_2 at high pressure is supplied through a hole in the tool and allowed to emerge from small channels under the tool tip as close as possible to the cutting edge. The expansion of the CO_2 lowers the temperature, and the tool close to the jet is kept below 0 °C. Improvements in tool life using this method have been confirmed but both CO_2 cooling and mist cooling are expensive to apply and have not been widely adopted in machine shop practice.

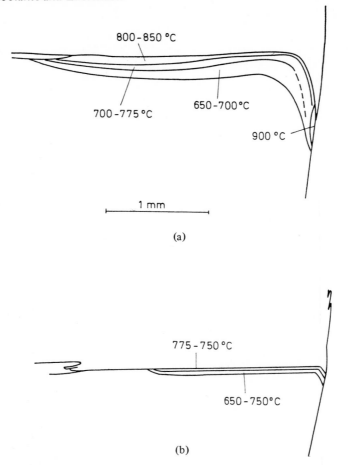

(a)

(b)

Figure 8.5   (a) Temperature contours derived from Figure 8.4a.   *(b) Temperature contours derived from* Figure 8.4b[2]

Lubricants

The term lubrication in relation to cutting fluids is used here to describe action by the fluid at the interface which reduces the tool forces and amount of heat generated, or improves surface finish and modifies the flow pattern around the cutting edge. Lubricants may improve the life of cutting tools but under some circumstances they cause an increase in the rate of wear.

Throughout the discussion of all aspects of metal cutting in this book, emphasis has been placed on the conditions of seizure at the tool/work interface. This emphasis has been necessary to secure proper appreciation of the most important problems of metal cutting, and also as a corrective to the framework of ideas conventionally used to treat the subject. This either entirely ignores, or greatly underestimates, the importance of this essential feature which distinguishes metal cutting from other metal working processes. Under conditions of seizure, however, lubricants apparently have no role to play, since the two surfaces are not in relative motion and, when metallurgically bonded, there is no possibility of access to the interface by an external lubricant whether solid, liquid or gaseous.

To understand the very important action of cutting lubricants, the emphasis must now be shifted to consideration of those areas of the tool/work interface which are not seized, and which have received little attention in previous chapters. Seizure may be avoided when the cutting speed is very low, as near the centre of a twist drill, or when cutting times are very short, as in the multi-toothed hobs used for gear-cutting. Under such conditions metal to metal contact may be very localised, and the tool and work surfaces may be largely separated by a very thin layer of the lubricant which acts to restrict the enlargement of the areas of contact. In other words, what are known as *'boundary lubrication'* conditions may exist, where a good lubricant, especially one with the extreme pressure additives chlorine and sulphur, will reduce cutting forces, reduce heat generation, and greatly improve surface finish.

Under the more normal conditions of cutting where seizure takes place, there is always a peripheral zone around the seized area where contact is partial and intermittent. This is shown diagrammatically in *Figure 3.13* for a simple turning tool, operating under conditions of seizure but without a built-up edge. In the peripheral region, the compressive stress forcing the two surfaces together is lower than in the region of seizure. The action of lubricants in this peripheral region, and the character of effective lubrication are shown most dramatically in experiments on metal cutting *in vacuo* carried out by Rowe and others.[3]

When cutting iron in a good vacuum (10^{-3} mbar) the area of contact between chip and tool increased greatly, the chip remaining seized to the tool for a much longer path than when cutting in air. This resulted in very high cutting forces and an extremely thick chip. The admission into the vacuum chamber of oxygen, even at a very low pressure, resulted in reduction of the contact area and tool forces to those found when cutting in air. The oxygen in air surrounding the tool under normal cutting conditions acts to restrict the spreading of small areas of metallic contact in the peripheral region into large scale seizure. It acts not to prevent seizure, but to restrict the area of seizure. The freshly generated metal surfaces on the underside of the chip and on the workpiece are very active chemically and are readily re-welded to the tool even after an initial separation. The role of oxygen in air or chlorine and sulphur in the lubricants is to combine with the new metal surfaces and reduce their activity and their affinity for the tool.

Thus air itself acts to some extent as a cutting lubricant, and if cutting were carried out in space, problems of high cutting forces and extensive seizure would be encountered. The action of air modifies the flow of the chip at its outer edge when cutting steel. In this region, oxygen from the air can penetrate some distance from the chip edge and act to prevent seizure locally, i.e., at the position *H–E* in the diagram, *Figure*

190

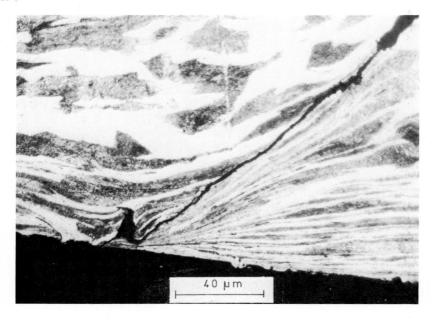

Figure 8.6   Section through steel chip near outer edge showing slip-stick action at interface in sliding wear region

Figure 8.7   Section through same steel chip as Figure 8.6, 0.5 mm from outer edge showing flow-zone characteristic of seizure at interface

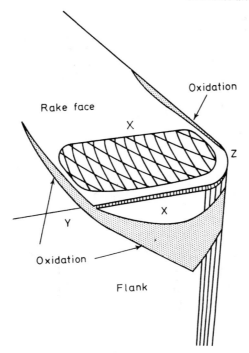

Rake face

Oxidation

X

Z

X

Y

Oxidation

Flank

Figure 8.8   Occurrence of oxide films on tools used to cut steel at high speed[4]

3.13.   Figure 3.3 shows scanning electron micrographs of a steel chip cut at 49 m min[-1] (150 ft/min) – note the segmented character of the outer edge.   Sections through this outer edge show a typical 'slip-stick' action, *Figure 8.6.*   The steel first sticks to the tool, but the presence of oxygen in this region restricts the bondings between tool and work material to small localised areas, the feed force becomes strong enough to break the local bonds, and a segment of the chip slides away across the tool surface.   The process is repeated, successive segments first sticking and then sliding away to form the segmented outer edge of the chip.   Oxygen is able to penetrate for only a short distance from the outer edge of the chip at this cutting speed, e.g., 0.25 mm (0.010 in) and, further inside the chip, seizure is continuous.   *Figure 8.7* is a section through the same chip at a distance of 0.50 mm (0.020 in) from the edge and shows a flow pattern typical of seizure.   The deep grooves sometimes worn in the tool at the position where the chip edge moves over the tool surface with this stick-slip action, *Figure 6.46,* are probably caused by chemical interaction of both tool and work materials with atmospheric oxygen.

Air contains 80% nitrogen and only 20% oxygen, and the nitrogen plays an important role in reducing oxidation of the tool when cutting steel and other metals at high speeds.   This can be demonstrated by a simple experiment.[4]   On those surfaces of the tool which exceed 400 °C during cutting, coloured oxide films are formed – the familiar 'temper colours'. If tools used for high speed cutting of steel are closely examined, two areas of the surface are seen to be completely free from the oxide films even though they were at high temperature during cutting. These regions, marked *X* in *Figure 8.8,* are on the clearance face just below the flank wear land, and on the rake face just beyond the worn area. The freedom from oxide films is not because these areas were cold during cutting – they were in fact closer to the heat source, and at a higher temperature than the adjacent oxidised surfaces. The explanation is that, during cutting, these parts of the tool surface form one face of a very fine crevice (the region just below *G* in *Figure 3.13* on the clearance face, and just beyond *C–D* on the rake face). The opposing face of this crevice is a freshly generated metal surface at high temperature, which combines with and eliminates all the oxygen available in this narrow space, leaving a pocket of nitrogen which acts to protect the tool surface from oxidation. Further down the clearance face and at the edges of the swarf more oxygen has access and the tool is coated with oxide films, although the temperature is lower. This is shown diagrammatically in *Figure 8.8*, and *Figure 8.9* is a photomicrograph of the clearance face of a carbide tool used to cut steel at high speed. The protective action of atmospheric nitrogen is eliminated if a jet of oxygen is directed into the clearance crevice. The tool is then very heavily oxidised in the region which was previously free from oxide films – compare *Figure 8.10* with *8.9*. Thus the oxygen of the air acts as a 'lubricant', reducing the area of seizure, while the nitrogen of the air modifies this action and largely prevents a serious problem of oxidation of the tools when cutting high melting-point metals at high speed, since carbide tools are oxidised very rapidly when exposed to air at temperatures of the order of 900 °C.

Figure 8.9 Photomicrograph of clearance face of carbide tool used to cut steel at high speeds in air[4]

Figure 8.10   As Figure 8.9 *but with jet of oxygen directed at clearance face*[4]

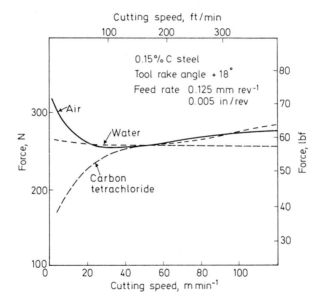

Figure 8.11   Influence of CCl_4 *and water as lubricants on tool forces in relation to cutting speed.   (After Rowe and Smart*[2] *)*

Other active elements, mainly chlorine and sulphur, introduced into cutting lubricants, probably act in a similar manner to oxygen.   Even when these are used in liquid lubricants they must act in the gaseous state at the interface because of the high temperature in this region. A flood of lubricant may be more effective than air because it eliminates the protective pockets of nitrogen and allows penetration of the active elements into the interfacial crevice, thus further restricting the area of seizure.   In this sense water probably acts as a lubricant as well as a coolant, penetrating between the tool and work surfaces and oxidising them to restrict seizure.

One of the most effective cutting lubricants is carbon tetrachloride (CCl_4), although this would not be considered as a lubricant at all under most sliding contact situations. Because of its toxic effects, CCl_4 cannot be used in industrial cutting. *Figure 8.11* shows the influence of CCl_4 and water on the tool forces when machining steel over a range of cutting speeds.[5] Because of their action in reducing contact area, active lubricants effectively reduce the tool forces. They are most efficient at low cutting speeds and have a relatively small effect at speeds over 30 m min⁻¹ (100 ft/min). By reducing forces, power consumption is reduced and temperatures lowered with consequent improvement in tool life. Not infrequently, however, the rate of tool wear is actually increased by the use of active lubricants.

Improved surface finish is a major objective of cutting lubricants. In this respect they are particularly effective at rather low cutting speeds and feed rates in the presence of a built-up edge. As an example, *Figure 8.12* shows the surface finish traces made using a 'Talysurf' on turned low carbon steel surfaces produced by cutting at 8 m min⁻¹ (25 ft/min) at a feed rate of 0.2 mm (0.008 in) per rev. both dry and using CCl_4 as a cutting fluid. The very great improvement in surface finish is caused by reduction in size of the built-up edge. Often, when cutting in air under such conditions, the built-up edge is very large compared with the feed, as shown diagrammatically in *Figure 8.13a* and in the photomicrograph *Figure 3.15*. The action of soluble oils, and of active lubricants is to reduce the built-up edge to a size commensurate with the feed, *Figure 8.13b*. Experiments have shown that the same result can be

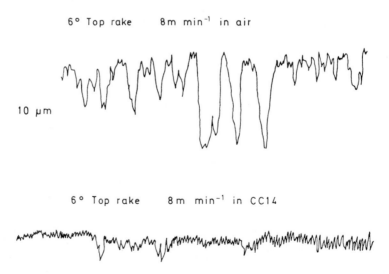

6° Top rake   8m min⁻¹ in air

10 μm

6° Top rake   8m min⁻¹ in CC14

Figure 8.12   'Talysurf' traces of steel surfaces cut dry and with CCl_4 as cutting lubricant

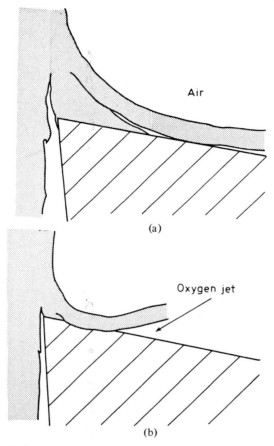

(a)

(b)

Figure 8.13 (a) Built-up edge when cutting dry. (b) Built-up edge when cutting with soluble oil lubricant[4]

achieved by the use of distilled water or of a jet of oxygen gas. This suggests that the stability of a large built-up edge, which consists of many layers of work material (*Figure 3.15*) is at least partially dependent on a protective pocket of nitrogen. When air is replaced by oxygen, water vapour, or some chemically active gas, which combines with fresh metal surfaces, adhesion is reduced between the layers of which the built-up edge is composed, and the stable built-up edge is much smaller.

In experiments on cutting steels at low speed and feed rate using carbide tools, the active lubricants which reduced the size of the built-up edge, also increased the rate of wear and changed its character.[3] The rate of flank wear was much increased and a shallow groove or crater was formed close to the cutting edge, as shown diagrammatically in *Figure 8.13b*. *Figure 8.14* shows the rake face wear on four tools after cutting *a* in air, *b* with a mineral oil, *c* with distilled water, and *d* with a jet of oxygen. The accelerated wear on the two latter tools appears

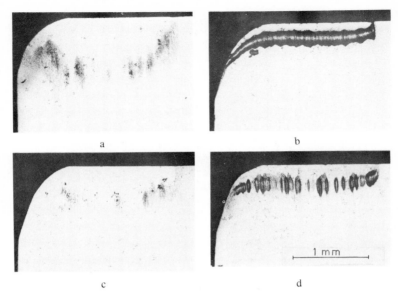

Figure 8.14   Wear on rake face of carbide tools when cutting steel at 30 m min⁻¹
(100 ft/min) at low feed rate⁴   (a) in air   (b) with mineral oil lubricant
(c) with water as lubricant   (d) with jet of oxygen directed under chip on rake face

to be the result of interaction of the tool and work materials with an active gas or liquid environment, while the mineral oil neither reduced the built-up edge, nor accelerated tool wear.   A large increase in wear rate has been observed also when a soluble oil cutting fluid was used during cutting of cast iron with carbide tools at medium and low speeds in the presence of a built-up edge.   Active lubricants can thus improve surface finish by modifying the flow of the work material around the cutting edge, although this may be at the expense of tool life.

Understanding of the action of coolants and lubricants in metal cutting is still at a rather primitive level.   The conditions are very complex, and practical men will wisely continue to rely on practical tests in development and selection of cutting fluids.   However, even the simple concepts put forward here may be of assistance in solving problems involving such a complex of requirements and process variables.

References

1.   MORTON, I.S., *I.S.I. Special Report 94,* 185 (1967)
2.   SMART, E.F. and TRENT, E.M., *Proc. 15th Int. Conf. M.T.D.R.,* 187 (1975)
3.   ROWE, G.W. and SMART, E.F., *Brit. J. App. Phys.,* **14,** 924 (1963)
4.   TRENT, E.M., *I.S.I. Special Report 94,* 77 (1967)
5.   CHILDS, T.H.C. and ROWE, G.W., *Reports on Progress in Physics,* **36,** 3, 225 (1973)

Bibliography

American Society for Metals Handbook, 8th Edition, Vol.3 'Machining' (1967)

ARMAREGO, E.J. and BROWN, R.H., *The Machining of Metals*, Prentice-Hall (1969)

BOOTHROYD, G., *Fundamentals of Metal Machining and Machine Tools*, McGraw-Hill (1975)

BROOKES, K.J.A., *World Directory and Handbook of Hard Metals*, Engineers' Digest Publication (1975)

KING, A.G. and WHEILDON, W.M., *Ceramics in Machining Processes*, Academic Press, New York

LOLADZE, T.N., *Wear of Cutting Tools*, Mashgiz, Moscow (1958) (in Russian)

PAYSON, P., *The Metallurgy of Tool Steels*, John Wiley & Sons, New York (1962)

ROBERTS, G.A., HAMAKER, J.C. and JOHNSON, A.R., *Tool Steels*, American Society for Metals (1962)

ROLT, L.T.C., *Tools for the Job*, B.T. Batsford, London (1968)

SCHWARZKOPF, P. and KIEFFER, R., *Cemented Carbides*, The Macmillan Co., New York (1960)

SHAW, Milton C., *Metal Cutting Principles*, M.I.T. Press, Cambridge, Mass (1957)

SWINEHART, Haldon J., Ed., *Cutting Tool Material Selection*, American Society of Tool and Manufacturing Engineers, Dearborn, Mich. (1968)

TAYLOR, F.W., 'On the Art of Cutting Metals', *Trans. A.S.M.E.*, **28**, 31 (1907)

ZOREV, N.N., *Metal Cutting Mechanics*, Pergamon Press (1966) (English translation)

INDEX